# Veterinary Clinical Pathology

# Veterinary Clinical Pathology

**Ganti A Sastry** GMVC BVSc MSc MS PhD (Kansas)
*Former* Principle and Professor of Pathology
College of Veterinary Science, Tirupati

CBSPD

**CBS Publishers & Distributors** Pvt Ltd

New Delhi • Bengaluru • Chennai • Kochi • Kolkata • Lucknow • Mumbai
Hyderabad • Jharkhand • Nagpur • Patna • Pune • Uttarakhand

# Veterinary Clinical Pathology



ISBN-13: 978-81-239-1183-0

**First Edition: 2000**
Reprint: 2001, 2004, 2005, 2006, 2007, 2009, 2014, 2016, 2019, 2022, **2024**



Published by **Satish Kumar Jain** and produced by **Varun Jain** for

**CBS Publishers & Distributors** Pvt Ltd

4819/XI Prahlad Street, 24 Ansari Road, Daryaganj, New Delhi 110 002, India.
Ph: 011-23289259, 23266861 Website: www.cbspd.com
e-mail: delhi@cbspd.com

*Corporate Office:* 204 FIE, Industrial Area, Patparganj, Delhi 110 092
Ph: 011-4934 4934 Fax: 011-4934 4935
e-mail: publishing@cbspd.com; publicity@cbspd.com

*Branches*

- **Bengaluru:** Seema House 2975, 17th Cross, KR Road, Banasankari 2nd Stage, Bengaluru 560 070, Karnataka, India
  Ph: +91-80-26771678/79 Fax: +91-80-26771680 e-mail: bangalore@cbspd.com
- **Chennai:** 7, Subbaraya Street, Shenoy Nagar, Chennai 600 030, Tamil Nadu, India
  Ph: +91-44-26680620, 26681266 Fax: +91-44-42032115 e-mail: chennai@cbspd.com
- **Kochi:** 42/1325, 1326, Power House Road, Opp KSEB, Power House, Ernakulum Kochi 682 018, Kerala, India
  Ph: +91-484-4059061-65,67 Fax: +91-484-4059065 e-mail: kochi@cbspd.com
- **Kolkata:** 147, Hind Ceramics Compound, 1st Floor, Nilgunj Road, Belghoria, Kolkata-700056, West Bengal, India
  Ph: +033-25633055, 033-25633056 e-mail: kolkata@cbspd.com
- **Lucknow:** Basement, Khushnuma Complex, 7 Meerabai Marg (Behind Jawahar Bhawan), Lucknow-226001, UP, India
  Ph: +0522-4000032 e-mail: tiwari.lucknow@cbspd.com
- **Mumbai:** PWD Shed, Gala no 25/26, Ramchandra Bhatt Marg, Next to JJ Hospital Gate no. 2, Opp. Union Bank of India, Noorbaug, Mumbai-400009, Maharashtra, India
  Ph: 022-66661880/89 e-mail: mumbai@cbspd.com

*Representatives*

- Hyderabad 0-9885175004
- Patna 0-9334159340
- Jharkhand 0-9811541605
- Pune 0-9664372571
- Nagpur 0-8692091830
- Uttarakhand 0-9716462459

*Printed at* Glorious Printers, Jhilmil Industrial Area, Delhi, India

*Dedicated to my late father*
*G.Y. Somayajulu Garu*

# PREFACE

Continued demand for my book by the student body had necessitated the production of this Sixth edition. It is no little satisfaction to the author to find that his book is felt needed and is found useful by most of the Veterinary students of the country.

In this edition some parts are revised and many facts and conditions have been added to bring the contents of the book uptodate. As such the number of pages had to be increased. To make the information more authentic. I have made a departure in this edition, by requesting Professors of other colleges as well as some of our own college, to subscribe chapters or conditions in which they have specilised. By this means the students get benefitted by thhe experience and experitise of these Professors. I am grateful for the following persons for contributing on the subjects noted against their names to this part :

*Ganti. A. Sastri*

# CONTRIBUTORS

The autor is greatly indebted to

Dr. S. Krishnaswamy, MVSC, DVPA & PH (DEN), PhD, professor and head, Department of Micorobiology, college of Veterinary Science, Tirupati, for revision of the Chapter on "Diagnostic Tests.

Dr. M. Mahendra, BVSC & AH. MS (Kansas), Associate porofessor & Head, Department of Pathology, College of Veterinary Science, Rajendranagar, Hyderabad, for his review of the chapter, "Hematology"

Dr. P. Naraisimha Rao, BVSC, PhD (Kansas), Associate Professor of Gynecology, College of Veterinary Science, Tirupathi, for the chapters, "Clinical Examination of Semen", "Pregnancy Diganosis" and "A Summary of infertility in cows".

Dr. P.K. Sriraman, BVSC, MSC, and Dr. A Suryanarayana Murthy, BVSC, MVSC, Assistanct Professors of Pathology, College of Veteirinary Science, Tirupathi for their help in the preparation of this book.

# CONTENTS


| *Chapter* | | *Page* |
|---|---|---|
| 1. | Hematology | 1 |
| 2. | Blood Chemistry | 25 |
| 3. | Urine | 51 |
| 4. | Mastitis Tests | 52 |
| 5. | Liver Function Tests | 56 |
| 6. | Cerebospinal fluid | 61 |
| 7. | Clinical Examination of Seman | 67 |
| 8. | Pregnancy diagnosis | 73 |
| | Ova in the faces of animals & fowls | |
| 9. | Clinical Parasitology | 78 |
| 10. | Diagnostic Tests | 82 |
| | Appendix I, A summary of causes of infertility in cows | 96 |
| | Appendix II, Materials to be sent for laboratory analysis | 99 |
| | Appendix III, Collection of specimens for laboratory examination | 100 |
| | Appendix IV, Materials to be submitted to the laboratory for diagnosis of a bacterial diseases | 103 |
| | A. Bacterial Diseases | |
| | B. Viral Diseases | |
| | C. Funglal Diseases | |
| | D. Prootozoal Diseases | |

# 1

# HEMATOLOGY

Functions of blood
Collection of blood
Anticoagulants used
- Oxalates
- EDTA

Erythrocyte counts
- Diluting fluids
- Procedure of counting

Packed cell Volume
Hemoglobin estimation
ESR
Mean Corpuscular Values
- MCV
- MCH
- MCHC

Reticulocytes
Leucocytes
- Diluting fluid
- Methods of counting
- Factors influencing total counts

Blood smears
- Staining
- Differential counts
- Factors influencing differential counts

Hematology of fowl
- Collection of blood
- Diluting fluid
- Counts

Clinical pathology is that branch of pathology that is utilised at the bedside of the patient. This involves the use of various laboratory procedures and the interpretation of the data obtained. Clinical pathology is of great help to the clinician in arriving at a correct diagnosis of the disease he is called upon to treat. Besides, it helps him in his correct prognosis of the condition. The laboratory tests also inform the clinician about the efficacy of his treatment.

## HEMATOLOGY

**The function of blood are:**

1. Transport of oxygen to the issues and carbon dioxide from tissues.
2. Transport of nutrients to the tissues.
3. Transport of products of metabolism for excretion through kidneys, lungs, liver, intestines and skin.
4. Control of water content of cells and interstitial fluid.
5. Regulation of body temperature by variation of its distribution.
6. Distribution of hormones for controlling body functions (Hormones in turn control the blood by action on hemopoietic organs).
7. Defence of body against bacterial invasion.
8. Immunity.

Since blood takes part either directly or indirectly in all biochemical processes of the body, it is but natural to expect alterations in it in disease. Hence examination of blood is a great adjunct in diagnosis of disease.

**Collection of blood :** Venous blood can be conveniently collected from the jugular vein in large animals. In dogs cephalic vein and the external saphenous vein may be used. The ear viens are used in the cat. In the pig blood can be collected from the anterior vena cava and in the rabbit and guinea-pig from the heart. Wing vein can be used in the fowl. In the case of rat and mouse blood is collected by sniping off a piece of the tail or from the heart direct.

For larger animals 5 ml of blood is quite sufficient for routine examinations. For small dogs and cats 2 ml of blood is sufficient.

Procedure for collection of blood: Note that for hematological work, the needle and syringe must be dry since presence of water hemolyses the red blood cells. The needle should in additon be sterile.

1. Clip the area from where the blood is to be drawn.
2. Apply tincture of iodine. Allow it to dry.
3. Raise the vein by pressure.
4. Use a large bore needle that has been sterilised and dried.
5. Insert the needle into vein, draw into the dry syring the required quantity of blood.
6. Disengage the needle, insert the syringe into the specimen tube (containing the anticoagulant) to very near its bottom and allow the blood to flow out without exerting any pressure on the piston. Immediately place the rubber stopper, place the tube on the plam of one hand and with the other make rotatory movement so that blood mixes with the anticoagulant.
7. Never shake the blood vigourously. Otherwise froth will form, red blood cells may be ruptured and erroneous results may be obtained.

**Anticoagulants to be used**

**Oxalates :** These are the cheapest. Oxalates combine with calcium and so coagulation is prevented. Ammonium oxalate swells the red cells while potassium oxalate shrinks them. So a mixture of these in the proportion of 60:40 is used to keep the red cells in their normal state. The mixture of these two salts is known as Heller and Paul's oxalate mixture. This consists of 1.2 gms of ammonium oxalate and 0.8 gms of potassium oxalate in 100 ml of distilled water, 0.5 ml of this solution is placed in specimen tubes, 1 1/2" x 3/4" and kept in a bacterialogical incubator overnight when the water evaporates leaving the salts. A rubber stopper is then provided to the specimen tube ( a bark cork should not be used because it will absorb plasma leaving blood more concentrated, thus yielding erroneous results). This oxalate mixture is sufficient for 5 ml of blood giving thus a concentration of 2 mg the of oxalate for 1 ml of blood. If smaller quantities of blood are utilised, proportionate quantity of the stock solution is pipetted into the specimen tube. For example, if only 2 ml of blood is to be collected, take 0.2 ml of the stock solution.

### Disadvantages

1. Since oxalate is toxic, blood containing this is not useful for blood transfusion.
2. Platelets tend to be clumped.
3. If the blood is kept for sometime, while blood cells develop artefacts.
4. Blood with oxalates is not useful for the estimation of NPN, BUN and lectrolytes.

**EDTA.** Ethylene diamine tetra acetate (a disodium salt is preferable). EDTA is a chelating agent, i.e. one that combines with calcium. This is an excellent anticoagulant. 1 mg of the powder for 1 ml of blood is used. If liquid is availabe 1 drop of a 10% solution is sufficient for 5 ml of blood. The solution is conveniently dispensed in plastic bottles which deliver a drop at a time when squeezed.

### Advantages

1. Clumping of platelets does not occur.
2. Normal staining is not affected.
3. At room temperature, blood can be preserved for 6 hours.
4. Suitable for total protein and BUN estimation.

### Disadvantages

1. Is not useful for blood transfusion.
2. Not suitable for estimation of calcium, magnesium, sodium, potassium and electrolytes.

### Erythrocyte Counts

For enumeration of red cells, we require the following :

1. A counting chamber with cover glasses.
2. A diluting pipette.

3. A diluting fluid.
4. A hand tally counter (if available)

Since the number of erythocytes in the blood are more than 4 to 5 millions they are too crowded in the counting chamber to make correct counts. So the blood has to be diluted sufficiently so that the red cells may lie singly in the counting chamber for correct counting. For this a diluting fluid is necessary. An ideal diluting fluid must have the following qualities:

1. It should not hemolyse the corpuscles.
2. It should promote their even distribution.
3. It should not bring about gross precipitation of proteins or other substances in the blood.
4. It should act as a fixative so that cells retain essentially their original shape.
5. It should prevent the growth of bacteria and moulds.

A fluid having the above qualities was formulated by the author and is now known as Ganti's fluid. It has the following composition :

| | | |
|---|---|---|
| Iodine crystals | 0.3 gms | |
| Potassium iodide | 0.4 gms | keep the bottle in |
| Sodium citrate | 2.0 gms | a dark place |
| Distilled water | 100 ml. | |

The advantages of this fluid are : (1) red cells are stained brown and so it is easy to count them on a faint yellow background. (2) cells are not clumped nor is their shape altered. (3) Setting of the cells in the counting chamber is quick.

Other diluting fluids that are being used by some workers are :

**Hayem's Fluid**

| | |
|---|---|
| Sodium chloride | 1.0 gm. |
| Sodium sulphate | 5.0 gms. |
| Mercuric chloride | 0.5 gms. |
| Distilled water | 200.0 ml. |

**Disadvantages are**

This fluid causes clumping of red cells. The mercuric chloride precipitates plasma proteins and red cells adhere to such such precipitated proteins and result in their speedy settling in the diluting pipettes. To overcome this, vigorous and continued shaking is necessary. It is difficult to count accurately the red cells in the clumps on the counting chamber. The red cells are also distorted in their shape.

## NORMAL DATA

| Specimen of Animals | Pulse | Respiration | Temperature F° | Temperature C° | Age of Puberty Months | Duration Days | Estrus period Average Days | Variation Days | Average period of gestation Days | First appearance of estrus after parturition. Days |
|---|---|---|---|---|---|---|---|---|---|---|
| Horse | 28-44 | 8-16 | 100.4-100.8 | 38-38.2 | 10-18 | 2-11 | 22 | 14-30 | 340 | 3-15 |
| Ox | 45-65 | 12-16 | 101.8-102.4 | 38.7-39.0 | 10-14 | 1-30 hours | 20 | Ass-374 16-24 Buffaloes 350 | 282 | 30-60 |
| Sheep | 70-80 | 12-20 | 101.3-05.8 | 38.4-41.0 | 6-12 | 1-3 | 21 | 10-24 | 148 | |
| Goat | 70-80 | 12-20 | 1009.103.8 | 38.2-39.9 | 6-12 | 2-3 | 21 | - | 148 | - |
| pig | 70-80 | 10-16 | 100.9-104.9 | 38.2-40.5 | 6-12 | 1-5 | 21 | 14-30 | 116 | 7-51 |
| Dog | 90-100 | 15-30 | 100.9-101.7 | 38.2-38.7 | 7-12 | 9-18 | Usually only | | 62-67 | Depends on |
| Cat | 100-120 | 16-32 | 100.6100.6 | 37.8-3.81 | - | 7-10 | once during each period | - | 55 | breed and environment |
| Camel | 35-50 | 5-12 | 95-101.5 | 35-38.6 | 36-60 | 5.0 | 23 | 22-24 | 395 | - |
| Elephant | 18 | - | 96-98 | 35.5-36.6 | - | - | - | - | 1921 month | - |
| Rabbit | 135 | 55 | 101.6 102.4 | 38.7-391 | 5-6 | Continuous | - | 5-16 | 31 | 35 |
| Mouse | 120 | - | 99.3 | 37.3 | 35-60 days | 10hrs | 4-5 | - | 10.20 | - |
| Guinea pig | 150 | 80 | | 99.6-38.9 | 60-80 days | 15hrs | 16.5 | 13-20 | 63-70 | Immediately |
| Rat | - | - | 99.5 | 37.5 | 100 days | 13-15hrs | 5 | - | 22.23 | - |
| Monkey | 100 | 20 | 101.0 38.3 | 5yrs | - | 28 | - | | 150-180 | - |

**Toisson's Fluid**

| | |
|---|---|
| Neutral glycerine | 3.75 ml |
| Sodium sulphate | 10.0 gms. |
| Sodium chloride | 1.25 gms. |
| Methyl violet 6B | 0.03 gms. |
| Distilled water | 2000 ml. |

**Disadvantages are**

1. This fluid allows the growth of bacteria and fungi.
2. It produces distortion of cells.
3. It requires longer time for complete settling of the cells in the counting chamber.
4. Since it is greasy, it is difficult for the pipettes to be quickly and efficiently cleaned.
5. Low values are obtained in anemias.

These fluids that distort the red cells, are not particularly useful for the erythrocytes of sheep and goats, in which the normal small cells are made to appear still smaller.

**Gower's Solution**

| | |
|---|---|
| Sodium sulphate | 12.5 gms |
| Glacial acetic acid | 33.3 ml |
| Distilled water | 200.0 ml |

**Disadvantages**

This solution also causes agglutination (clumping) of the red blood cells (especially those of sheep) and so makes correct counting difficult. Moreover, a red stain is left in the pipette and it is rather difficult to clean the pipette of this stain.

**Physiological Saline**

| | |
|---|---|
| Sodium chloride | 0.35 gm |
| Distilled water | 100 ml |

**Disadvantages**

This does not fix the cells and does not inhibit the growth of bacteria and so is not useful.

Hence it is opined that Ganti's fluid is better and gives more accurate results for the counting of erythrocytes in all animals.

**Procedure**

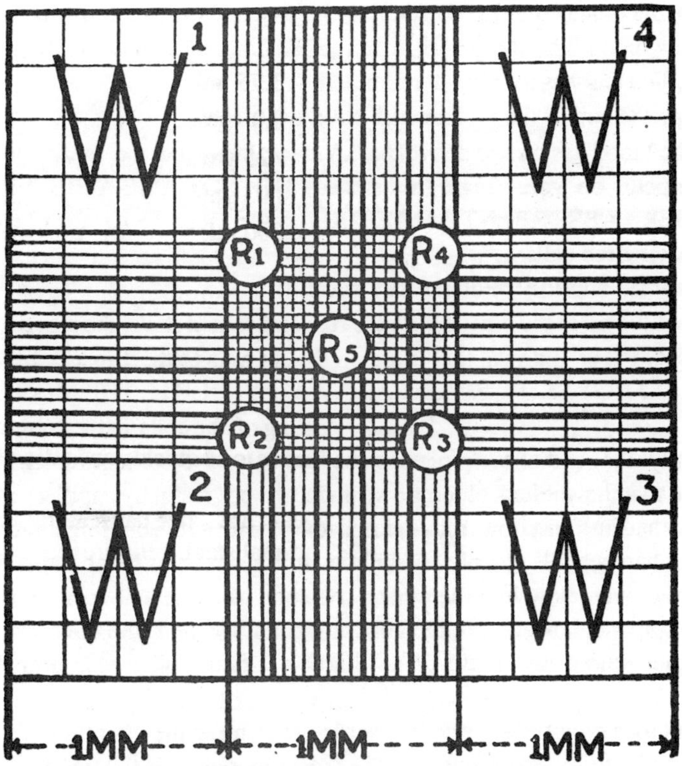


The rulings of the counting chamber denoting the areas for total erythrocyte (R) and leucocyte (W) counts.

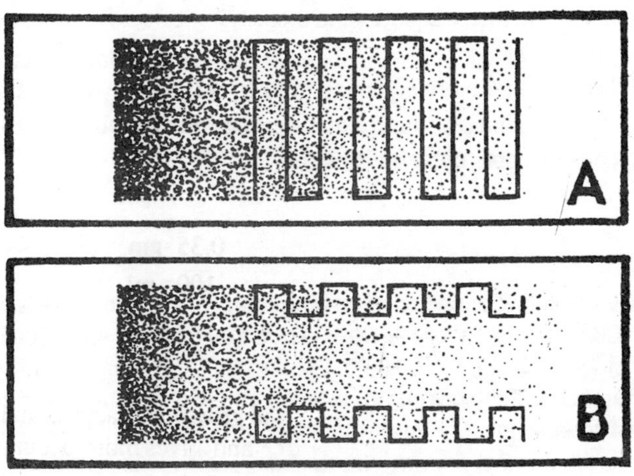


A-Method used for differential count.

B-Method used for examination of a blood smear for intraerythrocytic protozoa.

1. Mix the oxalated blood by rotating the specimen tube a few times.

2. Draw blood upto the 0.5 mark in the RBC pipette. (This is larger of the two pipettes supplied and has usually a red bead)

3. Wipe the end of the pipette with cotton or cloth.

4. If the blood has passed above the 0.5 mark, bring it to the mark by touching the end of the pipette on your finger (moistened) DO NOT USE A FILTER PAPER OR ANY OTHER ABSORBENT MATERIAL, since these being absorbent, plasma is absorbed leaving the cells.

5. Draw the diluting fluid upto 101 mark.

6. Close the tip of the pipette with the fore finger and shake the pipette vigourously.

7. Discard 2 or 3 drops (i.e. fluid in the stem up to the 1 mark) by twisting the rubber tube. DO NOT BLOW SINCE SALIVA MAY DILUTE THE MIXTURE IN THE PIPETTE.

8. Apply the end of the pipette at the junction of the counting chamber and the coverslip, so that the diluted blood flows under the coverslip by capillary attraction. To facilitate easy and uniform flow, rotate slowly the pipette with your finger intact. Note that the counting chamber fills completely without any air bubbles being trapped and that the fluid does not over flow into the moats. As soon as the chamber is completely filled, remove the pipette. (Note that the counting chamber and the coverglass should be perfectly clean without any grease or any particles of dust, otherwise, air bubbles will be trapped).

9. Wait for a minute for the cells to settle. Then bring the counting area at the centre which is divided into 16 small squares. (The area is marked R in the figure) into focus with the low power objective (2/3). Then change into the high power (1/6) so that one square divided into 16 squares and demarcated by tripple lines comes into field.

10. Count all cells in the 16 squares and those lying on two borders (leave those lying on the other two borders).

11. Count the cells in 4 more squares, marked R1, R2 etc. in the figure.

12. Add all the cells in five squares and multiply the sum by 10,000 to obtain the number of red cells in one cubic millimeter.

**Calculation**

The central area marked R is one square millimeter in area and this is divided into 400 small squares. Cells in 80 of these squares were counted i.e. those in 1/5 sq mm. The dilution of the blood is 200 (0.5 in 100). The depth of the fluid column (i.e. the depth between the chamber and the coverslip) is 1/10 mm. So the number of red cells in the 5 large or 80 small squares must be multiplied by 5x10x200 or 10,000. That is why if you add four zeros to the number for one c.m.m. Inherent error is $\pm$ 20%.

**Sources of Error**

1. Glassware not being dry.
2. Using a bark cork for the specimen tube.
3. Using absorbent material to reduce the column of blood to 0.5 mark in the pipette.
4. Not wiping the end of the pipette.
5. Using a pipette with a chipped end. (The pipette is calibrated to its tip. So if the end is chipped, then you are not taking correct quantity of the blood).
6. Not discarding the first 2 or 3 drops (the fluid in the stem is only the diluting fluid wihtout any blood and so this must be discarded).
7. Blowing with the mouth for discarding the first 2 or 3 drops.
8. Greasy chamber or chamber containing dust particles.
9. Air bubbles in counting chamber.
10. Flooding the chamber so that fluid runs into the moats.
11. Not allowing sufficient time for settling of the cells in the chamber.
12. Errors in counting.
13. Faulty pipette and counting chamber, with errors in calibration.
14. Inadequate shaking of the mixture in the pipette.
15. Contamination of the diluting fluid with yeast cells.
16. Drying of the fluid in the chamber before and during counting.

Increased number of erythrocytes can occur in :

Polycythemia.

Hemoconcetration : due to loss of fluid : diarrhoea, vomiting, excessive sweating, severe burns, polyuria, shock.

Decreased number occurs in Anemias, especially toxic, aplastic, homolytic and chronic hemorrhagic.

**The following characteristics are noticed in the erythrocytes of normal animals**

**Cattle :** Anisocytosis and crenation are common. The central pallor is only slight. Rouleaux is not formed.

**Horses :** Erythrocytes do not have central pallor. Rouleaux formation is common.

**Dogs :** The erythrocytes are large and uniform. The central pallor is notices as the eryhtrocyte is a biconcave disc. Crenation is not seen. A few blue colored cells - which are reticulocytes, are seen.

**Cats :** Central pallor is very slight. Crenation is common. Mild anisocytosis is seen. 1% of cells have H-J bodies. A few blue cells (reticulocytes, may be seen).

**Goats :** Among the domesticated animals, the erythrocytes of the goat are the smallest (average being 3.2 $\mu$) Rouleaux formation does not occur. Poikilocytosis is seen, erythrocytes being triangualr or oval.

**Buffaloe :** According to Purushottam & Mahendar (1963) the following are the normal values :

| | | | |
|---|---|---|---|
| RBCs — 5.8—6.2 millions/c mm | | ESR/1 hour | 55-60 |
| Hemoglobin —10.8—12.3 gms/100 ml | | Neutrophils | 43-45% |
| PCV —30—41 | | Lymphocytes | 45-48% |
| Total WBC –10.61 ± 2.0 | | Eosinophils | 3-4% |

Rouleaux formation occurs and hence is the high rate of ESR.

**Life Span of Erythrocytes Among Different Animals**

| | | | | |
|---|---|---|---|---|
| Horses : | 140-150 days | Sheep : | (adult) | 70-153 days |
| Cattle : | 160 days | Sheep : | (3 months) | 46 days |
| Goats : | 125 days | Dogs : | | 107-115 days |
| Cats : | 125 days | | | |

**Rouleaux Formation** : This is a grouping of erythrocytes resembling a pile of coins and is normally found in the enquine erythrocytes as well as in the erythrocytes of the buffaloe. In the erythrocytes of cat, pig and dog rouleaux formation may be found to some extent. But it is never found in cattle. If found in cattle, this denotes serious processes in the animals. The degree of rouleaux formation is directly proportional to the erythrocyte sedimentation rate, i.e. ESR is greatest in horses and buffaloes, moderate in dogs and least in cattle, sheep and goats in which no rouleaux formation is observed.

**Packed Cell Volume**

Packed Cell Volume (PCV) is the percentage of total volume occupied by packed erythrocytes when a given known volume of whole blood is centrifuged at a constant speed for a constant period of time. From this value obtained one can have an approximate idea of the number and size (and so volume) of the cells, because both these are altered in anemias, MCHC and MCV can be calculated from the PCV. For estimating PCV, EDTA of Heller & Paul's oxalate mixture or heparin is used as anticoagulant.

**Procedure**

A Wintrobe hematocrit is used. This is a tube with 3 mm bore and 10 cms in length. It is calibrated in centimeters. The calibration on the right starts with 0 from the bottom to 10 cms at the top. On the left, calibration starts from the top with 0. Both are subdivided into millimeters.

1. With a pasteur pippete (i.e. one with a long nozzle) fill the oxalated blood (which is mixed by a few rotations of the specimen tube) by inserting the nozzle into the tube so that it touches the bottom. While sqeezing the teat, slowly and gradually raise the pipette so that the Wintrobe tube gets completely filled to the top wihtout any air bubbles. See that the column of the blood is exactly upto 10 mark.
2. Put the tube into clinical centrifuge and rotate it at 3000 rpm (2260 G) for half an

hour. Then the red cells will be packed at the bottom with the plasma at top and leucocytes in between — the buffy coat

$$PCV = \frac{\text{Height of red cells in mm}}{\text{Total height of the column in mm}} \times 100$$

This is expressed as a percentage.

**Microhematocrit**

This method requires a special centrifuge, which may be costly but which is very accurate, and requires a very small quantity of blood. Capilary tubes, 7 cm X 1 mm are used.

**Procedure**

Mix the oxalated blood by rotation. Apply one end of the capillary tube to the surface of the blood, which rises into the capillary tube by capillary attraction and surface tension. When the capillary tube is 2/3 full, remove it and seal the end of it which is free of blood, by means of a small flame or by plasticine. Place the capillary tube into the groove of the centrifuge so that the sealed end is away from the centre. Fasten the cover of the centrifuge and then spin it for 2 minutes (12000 G). The reading is made with the help of a reader supplied with the instrument. The adventages of this method are: (1) very small quantity of blood is required, (2) shorter time is required for centrifugation and (3) greater accuracy since all the plasma is expressed and almost none is trapped between the red cells. The disadvantages are (1) we cannot estimate the Erythrocyte sedimentation rate (ESR) and (2) buffy coat is too thin a line to be easily interpreted.
The following data are elicited from a hematocrit.

1. Volume of the erythrocytes: A column shorter then normal indicated anemia whille one taller than normal hemoconcentration.
2. The leucocyte content of the blood: In the Wintrobe hematocrit, if the buffy coat is 1 mm or less, then roughly 0.1 mm is equivalent to 1000 leucocytes. Above 1 mm each 0.1 mm is equivalent to 2000 leucocytes, while in leucopenia, there may not be any buffy coat.
3. Nature of plasma. Normally it shoud be straw yellow. If deep yellow, icterus is denoted. In hemolytic anemia, it will be red. A turbid or opaque plasma indicates lipemia. This condition may be noticed (a) 2 or 3 hours after a meal, (b) in advanced cases of diabetes mellitus (c) in liver disease and (d) in hypothyrodism. The only errors that may probably arise are inadequate centrifugation and not using the oxalate mixture but only either ammonium oxalate or potassium oxalate.

**Note:** An anemic animal with severe hemoconcentration due to loss of fluids from the body (diarrhoea, vomition etc.) may have a PCV falling within normal limits. So while interpreting the PCV the water balance of the animal should be taken into consideration.

Usually $\frac{PCV}{6}$ = RBCs in millions.

## HEMOGLOBIN

There are many methods of estimation of hemoglobin but none are satisfactory. The cheapest and easiest that could be used for clinical purposes is the Sahli's acid hematin method.

### Method

1. Mix the oxalated blood in the specimen tube by rotation.
2. Place N/10, HCl in the diluting tube upto the mark 10 on the percentage side.
3. Draw the oxalated blood upto the mark in the pipette supplied. This indicates a volume of 20 c. mm.
4. Wipe the tip, and if the blood has overshot the mark, reduce it to the mark by tapping on the finger.
5. Introduce the pipette into the tube in which N/10 HCl is taken and immerse the pipette in the HCl. Allow the blood to flow out. If you keep the pipette vertical blood will flow out on its own accord. After all the blood has flown, raise the pipette slightly and such the acid and then blow it out gently without causing any air bubbles to disturb the acid and blood. Repeat this twice. Then draw distilled water into the pipette and then add this to the fluid in the tube. (These procedures are for rinsing the pipette to wash out any blood that may be sticking to its sides) NEVER BLOW STRONGLY TO BUBBLE THROUGH.
6. Shake the tube gently to mix the blood and acid.
7. Keep the tube in a dark place for 30 minutes. (or for length of time specifies by the marker)
8. At the end of this period, keep the tube in the comparator and then add distilled water drop by drop stirring with the glass stirrer supplied after each addition. This procedure is repeated till the color in the tube matches with the color of the standard in the comparator. While matching raise the stirrer form the solution. Restriction of the field of matching with the plam or a piece of paper enables one to appreciate better the difference in color between the standard and the solution.
9. When the colors match, then remove the tube from the comparator and read the level of the fluid in grams per 100 ml of blood.

### Sources of Error

1. Using a pipette with a chipped end.
2. Not taking correct quantity of blood in the pipette.
3. Taking too much of acid in the tube.
4. Error in matching (color blind people cannot match correctly).
5. Not keeping the mixture of acid and blood sufficiently long for the color to

develop.

Approximately, PCV /3 = Hb. in gms per 100 ml of blood.

Decreased hemoglobin values are seen in anemias. Cells in such conditions will be hypochromic.

## ERYTHROCYTE SEDIMENTATION RATE (ESR)

Due to weight, the erythrocytes gravitate towards the bottom if blood to which an anticoagulant is added, is kept verical in a hematocrit tube. The speed of fall in erythrocyte level in the tube differs in different animals. In cattle, sheep and goats, it is extremely slow, being less than 2 mm in 24 hours.

Among animals ESR may be of value only in the dog.

In certain destructive and inflammatory conditions, ESR may be accelerated. This may be due to some alteration in the physico-chemical properties of blood, in which the surface charge of the erythrocytes is altered, rendering them to aggregate together and form rouleaux. Rouleaux formation is accompanied by accelerated ESR. The larger the rouleaux formation, the speedier sedimentation occurs. There is also a rise in the fibrinogen and globulin content of the blood.

### Method

1. Use Heller & Paul's oxalate mixture or EDTA as anticoagulant.
2. Fill Wintrobe hematocrit tube to mark 0 with a long pasteur pipette from bottom upwards as explained earler, taking care to see that no air bubbles are present.
3. Keep the tube in a special stand, taking care to see that it is perfectly vertical. Even a slight enhances sedimentation.
4. Observe and measure the fall in the erythrocyte cloumn in millimeters from above (left side calibration) every 10 minutes.

### Sources of Error and Precautions to be Taken

1. The concentration of the oxalate must be exactly 2 mg/ml of blood. If it is higher, ESR may be slower.
2. Keep the hematocrit free from dirt. If it has been cleaned with ether or alcohol, dry it thoroughly, because dirt, alcohol and ether influence the rate of fall.
3. Keep the tube absolutely vertical.
4. See that no air bubbles are trapped while filling the tube since these influence the rate.
5. Conduct the test immediately after collection of blood, since too old a blood sample gives erroneous results.
6. See that hemolysis of blood does not occur, since it may modify the ESR.
7. See that the blood sample has normal temperature. Also see that the temperature of the place where the test is conducted is cool and between 22°C and 27° C. Higher temperature increases ESR.
8. See that the tube is kept in a place free from any disturbances or vibartions since these influence the ESR. DO NOT KEEP THE TUBE NEAR A RUNNING CENTRI-

## NORMAL VALUES (Figures in brackets indicate averages)

| Species animals | RBC10% cu. mm. | PVC % | Hb 2/100 ml | MCV μ 3 | MCH μμ g | MCHC % | RBC Diam μ | WBC $10^3$/ cu. mm. | N% | L% | M% | E% | B% |
|---|---|---|---|---|---|---|---|---|---|---|---|---|---|
| Cattle | 5-10 | 24-46 | 8-15 | 40-60 | 11-17 | 30-36 | 3.6-9.6 | 4-12 | 15-45 | 45-75 | 2-7 | 2-20 | 0-2 |
| | (7.0) | (35) | (11) | (52) | (14) | (33) | (5.8) | (18) | (28) | (58) | (4) | (9) | |
| Sheep | 8-16 | 24-50 | 8-15 | 23-48 | 9-12 | 321-38 | 3.5-6 | 4-12 | 10-50 | 40-75 | 0-6 | 0-10 | 0-3 |
| | (20-0) | (38) | (12) | (33) | (10.7) | (33) | (4.5) | (8) | (30) | (62) | (2.5) | (5) | |
| Goats | 8-18 | 9-38 | 8-14 | 15-30 | 10-126 | 35-42 | 3.2-3.9 | 4-13 | 10-59 | 40-75 | 0-6 | 0-10 | 0-2 |
| | (13) | (28) | (11) | (23) | | (38) | (3.2) | (9) | (30) | (62) | (2.5) | (5) | |
| Pigs | 5-8 | 32-50 | 10-16 | 50-68 | 17-23 | 30-34 | 4-8 | 11-22 | 28-47 | 39-62 | 2-10 | 1-11 | 0-2 |
| | (6.0) | (42) | (13) | (63) | (20) | (320 | | (16) | (37) | (53) | (5) | (3) | |
| Dogs | 5.5-9.5 | 37-55 | 12-18 | 60-77 | 19.0-25.0 | 32-36 | 6.7-7.3 | 6-17 | 60-77 | 12-30 | 3-10 | 2-10 | RARE |
| | (6.0) | (45) | (15) | (70) | (23.0) | (34) | (7.0) | (11.5) | (70) | (20) | (5) | (4) | |
| Cats | 5-10 | 25-45 | 8-15 | 39-55 | 13-17 | 30-36 | 5.4-6.5 | 5.5-19.5 | 35.75 | 20-55 | 1-4 | 2-12 | RARE |
| | (7.5) | (33) | (12) | (45) | (15) | (33) | (5.9) | (12.5) | (59) | (32) | (3) | (5) | |
| Horses | 5.5-9.5 | 24-44 | 8-14 | 34-58 | 12-18 | 31-37 | 58-8 | 5.5-12.5 | 35-75 | 15-50 | 2-10 | 2-12 | 0-3 |
| | (7.5) | (35) | (11.5) | (46) | (46) | (15) | (35) | (9.0) | (55) | (35) | (5) | (5) | |
| Elephants | 1.98-4.01 | 38.20 | 12-15.5 | 142-146 | 41-44 | 29-31 | 9.3 | 6.4-14 | 22-50 | 40-60 | 1.705 | 6-15 | 0.3-2 |
| | (2.81) | | (13.41) | (135.9) | | | | (10.16) | (36.53 | (51.73) | (2013) | (9.4) | (0.47) |
| Camel | 6.1-7.3 | 30.08 | 11.9 | 55 | 21-49 | 39-25 | 7.48x4.4 | 20.07 | 38.7 | 46 | 5.7 | 9.5 | 1 |
| | (7.26) | | | | | | | | | | | | |
| Rabbit | 4-7 | 30-50 | 8-15 | 68.0 | 24-7 | 35-0 | 6.7 | 6-12 | 17-52 | 42-80 | 5-8 | 0-3 | 0-5 |
| | (6) | (40) | (12) | | | | | (9) | | | | | |
| Guinea pig | 4-7 | 33-45 | 11-17 | 83 | 25 | 34 | 7.5 | 7-14 | 30-50 | 35-55 | 1-20 | 2-15 | 0-2 |
| | (6) | (40) | (14) | | | | | (10) | (42) | (45) | (8) | (5) | (0.5) |
| Rat | 7-10 | 35-45 | 12-18 | 57-65 | 18-22 | 31.7-35 | 6.3 | 5-23 | 8-40 | 50-80 | 2-7 | 0-4 | 0-2 |
| | (9) | (40) | (15) | | | | | (14) | (20) | (60) | (5) | (2) | (1) |
| Fowl | 2.18-4.12 | 27-42 | 6.83-11.3 | 127 | 37 | 29 | 12.2x7.3 | 20.35-33.3 | 29.5-37.3 | 48.9-58.4 | 9.7-10.2 | 1.7 | 0.7-2 |
| Man | 4-6 | 41-52 | 13.17 | 82-92 | 28-32 | 30-36 | 7.5 | 5-10 | 55-70 | 25-70 | 3-7 | 1-4 | 0-1 |

## NORMAL VALUES

| | SERUM | | | | BLOOD | | | | Plasma |
|---|---|---|---|---|---|---|---|---|---|
| Species of Animals | Ca. P. mg/100ml. | Mg. mg/100ml. | Total mg/100ml. mg/100ml. | Sugar Cholesterol | NPN mg/100ml. | Creatinine mg/100ml. | Urea mg/100ml. | Total niterogen | plasma proteins G/100ml. |
| Cattle | 9.4-12.2 (10.8) | 4-7.12 (5.56) | 2-5 | 50-230 | 35-55 | 20-40 | 1.0-2-07 | 6.0-27.0 | 7.56 |
| Sheep | 11.4 | 5.1-9.0 (5.21) | 2-56 | 64+12 | 35-74 | 20-38 | 1.2-1.93 | 8.0-20.0 | 5.18 |
| Goats | 10.7 | 3.0-11 | 2-3 | 55-200 | 45-60 | 30-44 | 0.9-1.82 | 13-28 | 6.25 |
| Pigs | 11.0-11.4 | 4-11 | 2-3 | 152-154 | 65-95 | 20-45 | 1.0-2.7 | 8-24 | 7.40 |
| Dogs | 9.8-11.6 | 2.2-4.0 | 2-1 | 125-250 | 70-100 | 17-38 | 1.0-1.7 | 10-20 | 6.1-7.8 |
| Cats | 8.22 | 5.3-7.6 (6.40) | 2-3 | 95-135 | 60-100 | - | 1.0-2.0 | 20-30 | 5.4-7 |
| Elephants | 6.90 | 3.2 | 2.2 | 93.6-116.7 | 42.73 | 45.46 | 1.62 | 16.19 | 10.72 |
| Horses | 12-13.4 (12.1) | 2.4-4.2 | 2.5 | 96.8 | 60-100 | 20-40 | 1.2-1.9 | 10-20 | 6.72 |
| Fowl | 9-12 | 4-8 | 5-9 | 58-94 | 125-275 | 20-35 | 1-2 | 0.4-1 | - |
| Man | 9-11 | 2-4 | 2-4 | 150-250 | 70-100 | 25-40 | 1-2 | 9-15 | 7.5 |

FUGE.

### Factors that Modify ESR

1. Age : ESR is higher in young and the old.
2. Pregnancy : It is higher in pregnancy.
3. Sex : In the human, is higher in females.
4. It is faster in Anemias.
5. It is higher in hypercholsterolemia.
6. It is higher in disease processes, in which there is inflammation.

ESR is of progonstic value since in a disease, a rate that is progressing towards normal is a favourable sign.

### Normal ESR in animals

| Species | In millimeters. Times 20 min | 1 hour |
|---|---|---|
| Elephant | — | 61-63 |
| Cattle | | |
| Sheep | 0 | 0 |
| Goats | | |
| Dog | — | 5 to 25 |
| Cat | — | 7 to 23 |
| Pig | — | 1 to 14 |
| Horse | 15 to 38 | — |

In dogs increased sedimentation rate if found in the following conditions : canine distemper, leptospirosis, salmon disease, chronic interstital nephristis, pneumonia, pleuricy, infectious canine hepatitis, myocarditis and myocardial degeneration, acute and subacute endocarditis, radiation injury, pregnancy.

## MEAN CORPUSCULAR VALUES

With the data available as to the total erythrocyte counts, hemoglobin values and PCV, we can estimate the volume of each erythrocyte and its hemoglobin content. This is helpful in determining what kind of anemia the animal is suffering from and also in helping us to select the correct therapy and to inform us if the treatment we are following is correct or not.

**Mean Corpuscular Volume (MCV)**

$$MCV = \frac{PCV}{\text{RBCs in millions}} \times 10 \text{ cubic microns. or } c\mu\mu$$

Example: If PCV of a sample is 30, RBC content is 5 millions.

$$MCV = \frac{30}{6} \times 10 = 60 \ c\mu\mu$$

**Mean Corpuscular Hemoglobin (MCH)**

$$MCH = \frac{\text{Hb in gms/100 ml}}{\text{RBCs in millions}} \times 10 \text{ micro microrgrams}$$

For example, if the Hb content is 12 gms./100 ml of blood and the animal has 4 millions of erythrocytes.

$$MCH = \frac{12}{4} \times 10 = 30 \ \mu\mu g.$$

**Mean Corpuscular Hemoglobin Concentration (MCHC)**

$$MCHC = \frac{\text{Hb in gms/100 ml}}{PCV} \times 100 \text{ per cent}$$

Example, if Hb is 12 gms% and PCV is 30, MCHC = 12 x 100/30 = 40%

MCV is helpful in the classification of anemias on morphological basis. Anemias are classified according to size as macrocytic, microcytic and normocytic. A higher MCV indicates a macrocytic anemia while a lower value is indicative of microcytic variety.

A decreased MCH denotes hypochromic anemia.

MCHC gives the quantity of Hb in each cell and also indicates the ratio of weight of Hb to the volume of cell. In most anemias this ratio is kept constant by the alteration of the cell volume and Hb content together. So MCHC is mostly normal. But in certain anemias there is reduction of Hb while the cell volume is normal; for example in iron deficiency MCHC is decreased.

There is compensatory mechanism in nature in providing the body with adequate amount of erythrocyte aeration surface. Even in the same species, a lower erythrocyte count is made up by an increase in the red cell diameter and hence there is increase in the MCV as well as increase in the MCH.

MCHC is equal in all species of animals, even in reptiles, amphibians and fishes.

## RETICULOCYTES

Reticulocytes are still immature red cells and are larger in size than the erythrocytes. These are not normally seen in the peripheral blood of cattle, sheep, goats and horses. In dogs 0.5 to 1% may be found while in pigs 2% may be encountered. Guinea pigs, and rabbits have to 2 to 4% normally.

Reticulocytes contain a fine network (or reticulum) which is believed to be RNA. this reticulum cannot be stained by ordinary stains, but only by surpa vital stains. Brilliant cresyl blue stains it.

Reticulocyte differs from the mature erythrocyte. In that (a) it is larger in size, (b) it is resistant to crenation, (c) does not participate in rouleaux formation, (d) it has a lower sp. gravity (e) is more resistant to hypotonic saline solutions, and (f) is polychromatophilic.

### Methods of Staining

1. Make a 1% solution of brilliant cresyal blue in saline.
2. In a small test tube mix a drop of the stain with a drop of the oxalated blood by rotation and stand for one minute and then make a smear of the mixture and stain with Leishman's stain. Count the number of reticulocytes among 100 erythrocytes.

The number of reticulocytes present is proportional to and is an index of erythropoietic activity, for the presence of a large number of reticulocytes indicates increased erythropoiesis.

Reticulocyte shower is the presence of a large number of reticulocytes suddenly in the peripheral blood and is an indication of a spurt in increased erythropoietic activity that occurs following acute hemorrhage.

The magnitude of reticulocyte response depends on:

(a) the size of stimulus, viz the degee of blood loss or blood destruction of deficiency. The more severe the anemia, the greater is the reticulocyte response.

(b) the quantity of blood building materials given, the portal of entry and the rate at which it is given. There is a certain maximum amount of material that can be given when the peak of response is slicited. Beyond this no further increase in response occurs. Liver extracts given parenterally is 30 to 60 times more effective than when given orally. A larger dose elicits an early and high peak while a prolonged administration of a low dose elicits a low and prolonged reticulocyte curve.

In chronic anemias as well as in hemolytic anemias there will be persistent reticulocytosis.

**Note:** Higher percentage of reticulocytes is found in hemolytic than in hemorrhagic anemias.

Reticulocytosis is absent in anemias that are due to absence of regeneration in bone marrow.

Reticulocytes are never seen in horses either in health or responding anemias.

### Use of Reticulocyte Count

(1) It is useful to detect anemias.

(2) It is useful to assess the response of an individual to therapy. 4th or 5th day following commencement of therapy retuculocytes appear in the peripheral blood and will disappear at the end of 2nd or 3rd week of therapy. If no response occurs one should then suspect some infection interfering with erythropoiesis or the therapy may be inadequate.

## LEUCOCYTES

The diluting fluid used for the enumeration of leucocytes has the following composition:

| | |
|---|---|
| Glacial acetic acid | 2 ml |
| 1% aqueous solution of gentian violet | 1 ml |
| Distilled water | 100 ml |

### Procedure

1. Draw blood upto 0.5 mark in the pipette which is smaller and has a white bead in the bulb.
2. Wipe the end, reduce the column of the blood upto 0.5 mark if it has gone above, taking care that you do not use an absorbent material to do so.
3. Then draw the diluting fluid upto 11 mark, shake well, discard two drops (i.e. the fluid in the stem) and then fill the chamber, taking the same precautions mentioned in connection with red cell counts. *Note:* The white cell pipette has a large bore and so if it is held vertical without closing the end with the finger all the fluid will flow out. So, keep your finger closing the end of the pipette.
4. Count the 4 large squares marked W. (each square has an area of 1 sq. mm.)
   Total WBC count = Total of the cells in 4 squares x 50/cm.
   The inherent error is $\pm$ 10%

### Factors that Influence Total Leucocyte Counts

1. **Age:** At birth total WBC count is high in calf and dog.
   In young pigs counts are low.
2. **Exercise:** Leucocytosis.
3. **Emotional states:** In fear, high leucocyte counts.
4. **Pregnancy:** Cow increased counts in later stages of pregnancy.
   Bitch leucocytosis nearer term
5. **Estrus:** Cows Higher counts on the day of estrus.
6. **Stage of digestion:** Dog. One hour after eating, counts higher.
   Pigs 3/4 hour after feed, higher counts.

Neutrophilic production is regulated by a "colony stimulating factor"(CSF) produced by bone marrow macrophages. Synthesis of CSF is stimulated by bacterial products in infections. Without CSF mitosis of stem cells does not occur.

Release of neutrophils from the bone marrow is promoted by a "leucocytosis inducing factor" (LIF) of the plasma. Concentration of LIF is increased in bacterial diseases by bacterial products. So it is that leucocytosis occurs in such diseases.

The neutrophilic number estimated in the peripheral blood does not represent the total neutrophils in the blood because the neutrophils moving sluggishly in the post capillary venules adhere to the vascular endothelium to make up the "marginal neutophil pool" (MNP). The neutrophils that move fast with erythrocytes constitute the "circulating neutrophil pool" (CNF).

In the dog MNP is equal to CNP while in the cat MNP is three times as much as CNP.

In states of fear, excercise excitement, parturition, rough handling, strange surroundings etc., there is leucocytosis (nutrophilia) since adrenaline that is liberated mobilises the MNP cells.

Similarly, there is leucocytosis (due to neutrophilia) in conditions where there is corticosteroid release as in states of stress--pain, anesthesia, trauma, manipulation in surgery, neoplasia etc. In this case stickiness of the neutrophils is reduced thereby increasing the CNP.

## THE BLOOD SMEAR

Slides should be clean. Make a smear immediately after blood is collected. The smear must be even and fairly thick, but not too thick. The smear should be dried at once by waving in the air. Do not blow by mouth as moisture in the breath may lake the red cells.

### Methods of Staining by Leishman's Method

1. Cover the blood smear with Leishman's stain — 1 minute.
2. Add double quantity of buffer, pH 6.6, and mix, by blowing. — Stain for 10 minutes.
3. Replace by distilled water, by flooding.
4. Differentiate in distilled water. (note that if you keep the water too long too much stain may be removed).
5. Dry at once. *Note.* The smear needs no fixing before staining. The methyl alcohol in the Leishman's stain fixes the smear.

**Note.** The buffer is prepared by dissolving 3.80 gms of disodium phosphate and 5.47 gms of monopotassium dihydrogen phosphate in 1000 ml of distilled water.

If the buffer is not available, neutral distilled water can be used.

Though for routine purposes, Leishman's stain is useful, but for bringing out the celluar details Leishman — Glemsa is better. The method used is as follows:

1. Flood the blood smear with Leishman's stain.
2. After half a minute add equal quantity of distilled water and mix by tilting the slide or blowing over the stain, taking care that nothing of the stain flows out.
3. After 5 minutes wash off the stain with distilled water.
4. Blot.

5. Flood the smear with diluted Giemsa's stain (Prepared by adding 1 drop of stock stain to one ml of buffer, 6.4 pH. The buffer is prepared by dissolving 6.63 gms. of mono potassium dihydrogen phosphate and 3.20 gms of disodium hydrogen phosphate in 1000 ml of distilled water.)

6. Allow to stain for 10 minutes, then wash off the stain with the buffer, differentiate with buffer for thirty seconds, dry and examine.

**Note:** The following points have to be borne in mind to obtain good staining of the smear:

1. Do not allow the stain to dry before it is diluted with the buffer. Otherwise stain particles will be precipiated.
2. Do not wash too long. Otherwise stain will be removed and the cells will not be stained.
3. The washing water or the buffer should not be too acidic. Or also the nuclei of the leucocytes will not be stained.
4. If the washing water or the buffer is too alkaline the cytoplasm of the leucocytes will not be stained.

Examine under oil immersion and count by Meander method, as shown in the figure. Avoid the tail end of the smear. Examine the edges fo the smear for intracellular parasites, as shown in the figure, since red cells containing the parasites, being lighter, tends to fall to the edges while making a smear. When intracellular parasites are present as in babesiosis, the cells though bigger become lighter. So, in the Wintrobe hematoerit, after centrifugation the infected cells are found just beneath the buffy coat, on top of the red cell column. So if a smear from this layer is made, it will be easier to make a diagnosis. Count at least 100 leucocytes (more will be better) and tabulate: neutrophils, lymphocytes, monocytes, eosinophils and basophils among these. While counting these cells, count also the immature neutrophils (myelocytes, juvenils and stabs—i.e. if there is a shift to the left).

While examining the blood smear for differential count the following characteristics of the erythrocytes, may also be noted :

1. Size: Whether the cells are of normal size or if they show any microcytes or macrocytes — anisocytosis.
2. Shape: Whether of normal shape or show poikilocytosis, or if Target cells are present.
3. Color: If normochromic or hypochromic.
4. Abnormal erythrocytes: Polychromotophilia, punctate basophilia, Howell-Jolly bodies, heinz bodies, Cabot rings, nucleated red blood cells (seen in anemias) and the presence of intracellular parasites.

**Factors that Influence the Differential Count**

1. **Age** Calf. During first week, low level of lymphocytes and higher neutro-

phils.
Swine. At birth 70% of lucocytes are neutrophils. In one week normal values are reached.
Dog. Lymphocytes are greater in number in young dogs.

2. **Breed** Cattle, sheep, goat and pig — Lymphocytes are always more in number than neutrophils.
Dogs and horse: Neutrophils are always more than lymphocytes.
3. **Exercise** During exercise there is neutrophilia (besides lucocytosis).
4. **Emotional states** There is neutrophilia in fear (besides the high leucocyte counts).

The following are the characteristics of the different types of white blood cells in different animals.

## NEUTROPHILS

Dogs: Between the nuclear lobes no true filaments are seen.
Cattle: Filament is seen in only one lobe.
Horses: The nuclear membrane appears jagged and this gives it a false multilobated appearance.
Pigs: Filaments are not present. The nucleus is coiled often.
Sheep: Nucleus much lobated.

## EOSINOPHILS

Horses: The eosinophilic granules are bright orange in color, are large and round-"Mulberry cell"—and obscure the cell outline.
Cattle: The granules are small, bright red, coarse and fill the cell (jewel-like)
Dogs: The granules are dull red, irregular in size and shape and do not fill the cytoplasm. Cytoplasmic vacuoles may be present.
Cats: The granules may be rod shaped.

## LYMPHOCYTES

Cattle: Two types, the small and large, are seen. The large one may be confused for a monocyte and has an indented nucleus which is light er stained than that of the small variety. The small lymphocyte has a compact nucleus with a rim of blue cytoplasm. Azurophile granules may be present in the cytoplasm.
Dogs: Only one type is seen, with a compact, round nucleus and scanty bluish cytoplasm.
Cats & horses: The lymphocytes resemble those of dogs.
Goats: Three varieties, small, medium and large, are seen.
Sheep: Only one type is seen. Azurophile granules are seen in the cytoplasm.

## MONOCYTES

Horses: The nucleus is typically kidney-bean shaped.
Cattle: The nucleus is less compact than that of the lymphocyte and is of varying

shapes. Cytoplasm is vacuolated.

Dog: Sometimes the kedney-bean shape of the nucleus may be seen. Vacuolation may be seen in the cytoplasm and at times in the nucleus also. Nucleus is of varying shapes.

Sheep and goats: The nucleus is amoeboid or three pronged.

**Normal neutrophil : lymphocyte ratio**

| | | | |
|---|---|---|---|
| Cattle | 28:58 | Sheep | 30:60 |
| Dogs | 70:20 | Horses | 55:35 |
| Cats | 60:30 | Pigs | 35:50 |

**Hematology of the Fowl**

(The following information is from the M.Sc., thesis of late Dr. K.T. K. Nambiar of the Madras Veterinary College, entitled ''Studies on the hematology of the domestic fowl''.

**Collection of the blood sample** Insert a sterile 18 to 20 gauge hypodermic needle into the wing vein and draw 2 ml of blood through a dry syringe. Heller and Paul's oxalate mixture can be used as an anticoagulant.

**Diluting fluid** For erythrocyte counts, diluting fluid described already can be used. But for leucocyte counts the ordinary white blood cell diluting fluid is not useful since, though the erythrocytes may be lysed, the nuclei are left and appear prominently and so it is difficult to distinguish the leucocytes from them. So to overcome this difficulty, differential stains are used so that the erythrocytes and leucocytes stain differently and can be counted easily. The following diluting fluid can be used with which both erythrocytes and leucocytes can be counted. Make the following stock solution.

A...Sodium citrate 2%.
B...0.1% Gentain violet in Ringer's solution.
C...0.1% Brilliant cresyl blue in Ringer's solution.
D...Neutral formalin......

The above 4 solutions must be made and kept separtely as stock solutions. From these the working solution is made daily before making the counts as follows:

A.......1 ml
B.......2 ml
C.......1 ml
D.......3 drops.

Mix and filter.

Ringer's solution has the following composition:

Sodium chloride ... 0.7 gms
Sodium bicarbonate ... 0.03 gm
Potassium chloride ... 0.026 gm

Calcium chloride .... 0.003 gm
Distilled water ... 100 ml

With the above solution, erythrocytes take a very light blue color and clearly seen. Granulocytes are prominent. Thrombocytes are also clearly visible and uniformly distributed and can be easily counted. Moreover, with the above fluid the heterophils, monocytes and lymphocytes can be identified easily.

**Counts**

Take a red blood cell diluting pipette and draw the oxalated blood upto 0.5 mark and then draw the diluting fluid upto the 101 mark, thus making a 1 in 200 dilution. Mix well. As for the mammalian blood, count the erythrocytes in 80 small squares. Calculate the number of erythrocytes by adding 4 zeros to the total number you got in the counting in the 80 small squares.

The same blood mixture and the same counting chamber can be used for leucocyte counts also. Since the number of leucocytes are far greater in number in fowls than in the mammalian blood, a 1 in 200 dilution is indicated.

Count the leucocytes on the 9 large squares having a total are of 9 sq. mm. and count the average for 1 sq. mm. The figure obtained, multiplied by 2000 gives the total leucocytes in one cu. mm. of blood.

Hemoglobin content, Packed cell volume and Erythrocyte sedimentation rates are estimated as for the mammalian blood. In making a differential count, the cells are classified as heterophils, eosinophils, basophils, lymphocytes and monocytes.

## CHARACTERISTICS OF THE BLOOD CELLS OF THE FOWL

**Erythrocytes** The erythrocytes of the fowl have an oval shape. They are nucleated. Certain amount of poikilocytosis may be seen normally. Occasionally rounded forms may be seen and these are considered to be younger and less mature, having rounded nuclei which are less thick.

Usually the male birds have higher values for erythrocyte counts and hemoglobin values.

## LEUCOCYTES

**Heterophils** These cells which are not perfectly round, have acidophilic crystalline granules in the cytoplasm which is colorless and clear. In the chicken these granules are rod or spindle shaped. In the ducks, on the other hand, these are round. The lobated nucleus has two or three lobes. Occasionally a single lobe may be seen.

**Eosinophils** Having the same size as that of the heterophils, eosinophils have in their cytoplasm round, dull red granules. The granules in the duck are rod shaped. The cytoplasm has a slightly bluish—gray color and the nucleus is bilobed mostly.

**Basophil** This also is of the same size as the heterophil but its nucleus which is lobated, is maked by the dark staining basophilic granules.

**Monocytes** Sometimes these are difficult to be distinguished from the larger lymphocytes, but the monocytes are larger in size and the cytoplasm has a blue-gray color. The outline of the nucleus is irregular. Males appear to have more number of monocytes than the females.

**Lymphocytes** These are most numerous in the fowl blood. Wide variation in size and shape is noticed. The nucleus is usually round or sometimes may be indented on one side. The cytoplasm in the smaller variety is just a narrow rim taking a faint bluish stain. In the larger varieties, cytoplasm is more. Sometimes non-specific azure granules may be noticed. The number of lymphocytes, may be slightly more in the female than in the male. They are also more in the young than in the adult.

Blood Glucose
- Estimation
- Hyperglycemia
- Hypoglycemia

Hypermagnesemia
hypomagnesemia
Serum alkaline phosphatase
- Estimation

# 2

# BLOOD CHEMISTRY

Serum Calcium
Estimation
Hypercalcemia
Hypocalcemia
Serum Inorganic phosphorus
Estimation
Hyperphosphatemia
Hypophosphatemia
Serum Magnesium
Estimation
Increased values
SGOT and SGPT
Estimation
SGOT activity
SGPT activity
Increased values
Serum Cholesterol
Estimation
Hypercholesteremia
Hypocholesteremia

## BLOOD GLUCOSE

### (Neison - Somogyi Method)

**Principle:** Blood is deproteinized by a Zinchydroxide barium sulphate procedure which gives a filtrate containing practically no reducing substances other than glucose. The zink barium filtrate is heated with alkaline copper reagent and then treated with a special arseno molybdate color reagent. The color developed is compared with that obtained from a known amount of glucose.

**Procedure:** To 1 ml of blood in a 50 ml. flask add 9.5 ml. of barium hydrroxide with mixing. Add 9.5 ml. zinc sulphate solution mixing by rotation. Shake and filter. To 0.5 ml. of filtrate in a 10 ml. calibrated tube add 1 ml. alkaline copper reagent, mix by tapping and keep for 20 minutes in a boiling water bath. Cool for one minute and add 1 ml. arsenomolybdate reagent, mix and make up to 10 ml. with distilled water. Prepare a blank and standard (0.025 mg. glucose/0.5 ml.) in the same way. Read the color at 540 mu.

**Calculations:** (For colorimetric measurement)

$$\frac{\text{Reading of unknown}}{\text{Reading of standard}} \times \text{mg. glucose in standard} \times \frac{100}{0.025} = \left[ \begin{array}{c}\text{mg. of glucose} \\ \text{per 100 ml} \\ \text{of blood.}\end{array} \right.$$

## REAGENTS

1. **Barium hydroxide solution:** Dissolve 90 Gms of $Ba(OH)_2$, $8H_2O$ in distilled water and dilute to 2000 ml in a graduated cylinder. Filter if cloudy. Store in well stoppered containers filled to capacity.

2. **Zinc sulphate solution:** Dissolve 100 gms of zinc sulphate in distilled water, dilute to 2000 ml in a graduated cylinder and mix. Stable indefinitely.

It is most important that these two solutions exactly neutralize each other. Check by measuring 10 ml of Zinc sulphate solution, into a flask, diluting with 50 ml distilled water, add 4 drops of phenolphthalein solution and titrate with barium hydroxide soln. slowly and with constants shaking until 1 drop of alkali turns the mixture pink. It requires more or less 10 + 0.05 ml of barium hydroxide. If not, dilute one or the other solution as required and check again. Continue until the two solns are exactly equivalent. Protect the barium hydroxide solns in dispensing bottle with soda lime tube in the stopper and test each soln Occasionally by making trial filtrate on blood. Satisfactory solns. give a clear filterate which filters rapidly and show little tendency to foam.

3. **Alkaline copper reagent solution A:** Dissolve 50 gms of anhydrous sodium carbonate; 50 gms. of Rochelle salt (pot sod. tartarate), 40 gms. Sodium bicarbonate and 400 gms of anhydrous sodium sulphate in about 1600 ml of distilled water and dilute to 2 liters. Mix and filter if not clear. Store at room temperature. If a sediment forms in a few days, filter again.

4. **Solution B:** Dissolve 150 gms Copper sulphate ($CuSO_4.5H_2O$) in distilled water and dilute to 1 litre. Add 0.5 ml of conc. sulphuric acid and mix.

5. **Alkaline copper reagent:** On the day it is to be used, place 4 ml of solution B in 100 ml graduated cylinder, dilute to 100 ml with solution A and mix.

6. **Arsenomolybdate color reagent:** Dissolve 100 gms of ammonium molybdate in 1800 ml of distilled water. Add 84 ml of concentrated sulphuric acid, with stirring. Dissolve 12 gms of disodium orthoarsenate in 100 ml of distilled water, and add it with stirring to the acidified molybdate solution. Place the mixture in an incubator at 37° C. for 1 or 2 days. Store in a glass stoppered brown bottle. This solution should be stable indefinitely.

7. **Standard glucose solution:** Prepare a stock standard by dissolving 1 gm of highest purity anhydrous glucose in about 10 to 15 ml of 0.2% Benzoic acid solution, transfering quantitatively to 100 ml volumetric flask, diluting to the mark with the benzoic

acid solution and mixing. This solution is stable indefinitely and contains 10 mgs of glucose per ml.

**8. Working standard** (0.025 mg glucose per 0.5 ml): Dilute 1 ml of the stock standard to 200 ml with 0.2% benzoic acid solution and mix.

Physiologically, blood glucose level is increased after feeding in the monogastric animals and during excitement. Decreased levels are found during strenuous and prolonged exercise and in starvation.

The action of epinephrin, noradrenaline, glucocorticoids, STH and glucagon in elevation of glucose levels and insulin in lowering of the blood glucose level have been already discussed.

**Conditions in which Hyperglycemia May Occur**

1. Diabetes mellitus in the dog, cat, cow and sheep.
   This may arise due to
   (a) Pituitary neoplasms
   (b) Adrenal hyperplasia
   (c) Pregnancy toxemia
   (d) Acute pancreatic necrosis

2. Convulsiions as in
   (a) Eclampsia
   (b) Intracranial trauma
   (c) Epilepsy
   (d) Tetany
   ] Probably due to anoxia and increased epinehrine secretion

3. Transient conditions
   (a) Exposure to cold,
   (b) Following general ancsthesia,
   (c) Administration of morphine,
   (d) Injection of epinephrine,
   (e) Intravenous glucose injection,
   (f) Feeding large quantities of carbohydrates and
   (g) Excitement.

4. Hyperthyrodism—in this there is increased glucose production by the liver in excess of utilisation.
5. Chronic liver disease.
6. Chronic nephritis.

**Conditions in which Hypoglycemia May Occur**

1. Starvation
2. Hypopituitarism

3. Hypothyroidism
4. Hyperinsulinism (a) Tumor of pancreas
   (b) Excess of insulin in therapy.
5. Adrenal cortical insufficienty
6. Severe exertion—-glucose utilisation is increased and liver glycogen is depleted.
7. Ketosis in cattle and sheep.
8. Hypoglycemia of baby pigs. (manifested by convulsions, weakness, coma, and death) Blood glucose level is 40 mg%.

## CALCIUM

### Estimations of Serum Calcium

In animals serum calcium level is 9 to 11 mg/100 ml. Calcium estimation of serum is indicated in the following conditions : (1) bone diseases. (2) convulsions, (3) Parathyroid disorders.

**Principles of estimation:** Ferro—Hem method. Calcium is precipitated by chloranilic acid. The insoluble calcuim chloranilate is washed free of excess chloranilic acid with 50% isopropyl alchol and then dissolved in the tetrasodium salt of ethylaenediamine tetra acetic acid (EDTA), forming a pink soluble sodium chloranilate. The stable pink soultion is read at a wavelength of 520 m $\mu$.

### Reagents Required

**Chloranilic acid:** Dissolve 4 gm of sodium hydroxide in 600 to 700 ml of distilled water. Add 11 gm chloranilic acid and suffcient distilled water to make 1 liter. Agitate for 10 minutes or until all the acid is dissolved. The solution should have a pH between 3.5 and 7.0. If the pH is higher than 7.0. add additional 1 gm quantities of chloranilic acid to reduce the pH below 7.0. Let stand overnight. Fiter through a Buchener funnel containing a fine pad.

**Isopropyl alcohol:** Dilute 50 ml of isopropyl alcohol to 100 ml with distilled water.

**Tetrasodium ethylenediaminetetraacetate:** Prepare a 5% solution by dissolving 5 gm of the salt in distilled water and adjust the volume to 100 ml.

**Calcium standard:** Weigh exactly 0.2497 gm of reagent grade calcium carbonate and transfer to 1 liter vloumetric flask. Add 2 ml of dilute hydrochloric acid (1 ml. of conc. HCI plus 8 ml distilled water). Allow to stand until all the carbonate has dissolved. Add dissolved. Add disstilled water to the mark and mix. This is equivalent to 10 mg of calcium per 100 ml.

**Procedure**

1. Pipette 2ml of serum and 2ml of standard into heavy walled centrifuge tubes. Mark them suitably.
2. To each tube add 1 ml of the chloranilic acid solution agitating the tubes constantly.
3. Allow the tubes to stand for at least 30 minutes
4. Centrifuge at approximately 1800 rom. for 10 minutes.
5. Decant the supernatant and allow the tubes to drain for 2 or 3 minutes on some absorbent paper such as filter paper.
6. Wipe the mouth of the tube dry with cotton gauze.
7. Wash the precipitate with 6 or 7 ml of 50% isopropyl alcohol. A fine stream from a polyethylene wash bottle should be used. Break up the sediment and resuspend it.
8. Cetrifuge and drain as before.
9. Add 2 drops of distilled water to each packed precipitate.
10. Break-up precipitate by striking the bottom of the tube sharply against the palm of the hand until the mat breaks loose and the precipitate is suspended in water. if this is not successful a dry broom stick may be used to loosen the precipitate.
11. To each tube add 6 ml. of EDTA soultion.
12. Stopper the tubes and invert them several times until the precipitate is compeletely dissolved. Do not agitate vigourously.
13. Transfer the pink solution to cuvgettes and read at 520 mu after adjusting to zero optical density with distilled water as blank. The color is stable for 5 days.

For greater accuracy use the same cuvette for all the readings.

**Calculation**

$$\frac{\text{Density of unknown}}{\text{Density of standard}} \times 10 = \text{mg of calcium per 100 ml of unknown}$$

**Note**

1. Do not use standard contaminated with mold growth.
2. Erroneous results will ensue if you do not keep the tubes for at least 30 minutes after adding the chloranilic acid in step 2 above.
3. Use a centrifuge having free swinging cups. Angle head centrifuge is not useful since with this the precipitate does not form a tight buton.
4. In step 12 see that all the precipitate is dissolved.

**Hypercalcemia may be found in**

1. Feeding large quantities of vitamin D.
2. Parathyroid adenoma (Primary hyperparathyroidism).
3. Renal disease culminating in secondary hyperparathyroidism.
4. Neoplasms of the bone; Primary—Ewing's tumor
   secondary—metastases.

5. Hyperproteinemia.

**Hypocalcemia may be found in**

1. Hypoparathyroidism.
2. Hypoproteinemia—due to starvation. Here the bound calcium is decreased.
3. Deficiency of Vitamin D in feed and calcium intake.
4. Ketosis—there is increased calcium excretion in urine in this condition.
5. Milk fever.
6. Rickets.
7. Eclampsia.

## INORGANIC PHOSPHORUS

**Method of Fiske and Subba Row** Principle. The proteins of the serum are precipitated with trichloracetic acid. When treated with an acid molybdate solution, the phosphate present in the protein-free filtrate forms phosphomolybdic acid. This is reduced to phosphomolybdous acid by 1,2,4- aminonaptholsulphonic acid reagent, forming a blue color, the intensity of which is proportional to the amount of phosphate present.

**Reagents Required**

1. **10 per cent trichloroacetic acid** Dissolve 10 gm of high-reagent grade trichloroacetic acid crystals in distilled water and make up the volume to 100 ml.

2. **5 per cent trichoroacetic acid** Dissolve 5 gm of trichloroacetic acid in distilled water and make up to 100 ml.

3. **10 Normal sulphuric acid** Carefully add, slowly 450 ml of concentrated sulphuric acid (sp. gr. 1.84) to 1300 ml of distilled water. This can be checked by diluting 10 ml to 100 ml and titrating aginst a one N. sodium hydroxide. If necessary adjust the suphuric acid. Store in glass stoppered Pyrex bottles.

4. **Molybdate I. solution** Dissolve 12.5 gm of ammoniumj molybdate in 100 ml of distilled water in a 500 ml volumetric flask. Add 250 ml of 10 N sulphuric acid and dilute to the mark with the distilled water.

5. **Molybdate II, solution** Prepare the previous solution but with only 150 ml of 10 N sulphuric acid.

6. **15 per cent sodium bisulphite solution** Dissolve 75 gm of reagent-grade sodium bisulphite in distilled water and dilute to 500 ml Should the solution show slight turbidity. keep for 2 or 3 days and use the supernatant fluid. keep well stoppered.

7. **20 per cent sodium sulphitesolution** Dissolve 20 gm of reagent-grade anhydrous sodium suphite in distilled water and dilute to 100 ml filter if necessary and keep well stoppered. If anhydrous salt is not available, use the crystalline salt, $Na_2 SO_3 . 7H_2 O$) but disolve 40 gm of the salt.

8. **Aminonaphtholsulphonic acid Reagent** Place 195 ml of the 15 percent sodium bisulphite solution in a suitable glass stoppered container and add 0.5 gm of 1,2,4-

aminonaphtholsphonic acid and 5 ml of the sodium sulphite solution. Stopper and shake vigourously until the powder is dissolved. If powder dose not dissolve add 1 ml of sulphite and shake, keep in a dark glass-stoppered bottle in a refrigerator. Keeps for 4 weeks.

9. **Standard phosphate solution** In a mortar grind finely reagent grade potassium dihydrogen phosphate and dry over a vacuum chloride dessicator for 3 or 4 days. Weigh 0.3514 gm with an analytical balance and dissolve in distilled water. Transfer to a 1 liter flask and add 10 ml of 10 N sulphuric acid. Dilute carefully to the mark, mix thoroughly. This solution keeps indefinitely if well stoppered. 5 ml = 0.4 mg of phosphorus.

10. **For photoelectric colorimeter:** Pipette 5 ml of the standard phosphate solution described above into a 50 ml volumetric flask and dilute to the mark with the 5 percent trichloroacetic acid solution 1 ml. contains 0.008 mg of phosphorus.

**Procedure with Visual Colorimeter**

1. Take in a 50 ml conical falsk 8 ml of the 10 percent trichloroacetic acid and add 2 ml of the serum while gently rotating the flask. Stopper with a rubber stopper and shake vigourously and filter at once through a dry ashless quantitative filter paper into a dry clean beaker. Note: There should not be any hemolysis.

2. Pipette 5 ml of the filtrate into 10 ml volumetric flask and add 1 ml of the molybdate II solution.

3. Pipette 5 ml of the standard phosphate solution (5 ml = 0.4mg P) into a 100 ml volumetric flask. Add approximately 60 ml. of distilled water and 10 ml. of molybdate 1 solution.

4. Add 0.5 ml. of the aminonaphtholsulphonic acid reagent to the 10 ml flask, dilute to 10 ml mark and mix. Immediately add 4 ml. of the aminonaphtholsulphonic acid reagent to the 100 ml flask dilute with distilled water to the mark and mix. Keep for 5 minutes.

5. Compare the standard against itself before reading the unknown. If the unknown is darker, keep for 10 minutes longer and read again.

**Calculation:** Note that the standard contains 0.4 mg of P for 100 ml. So 10 ml of this contains 0.04 mg P.

$$\text{So, mg, of P in 100 ml.} = \frac{\text{Reading of standard}}{\text{Reading of unknown}} \times 0.04 \times \frac{100}{1}$$

**Procedure with Photoelectric Colorimeter**

1. Add 1 ml of serum to 9 ml of 5 percent trichloroacetic acid in a flask, mix well. Let stand for 5 minutes.

2. In a 10 ml volumetric flask take 5 ml of the filtrate (which is equivalent to 0.5 ml of serum), add 1 ml of myolybdate II solution and 0.4 ml, of the aminonaphtholsuphonic acid reagent and then dilute to mark with distilled water. Mix well. Let stand for 5 minutes.

3. In a 10 ml volumetric flask, to 5 ml of diluted phosphate standard solution add the same quantitiies of reagents used in the unknown above and make up to the mark with distilled water. Mix well and stand for 5 minutes and read in the colorimeter.

4. A reagent blank is made by taking 5 ml of the 5 percent trichloroacetic acid and adding the same reagents as in the unkown and making upto 10 ml and read in the colorimeter.

5. Use a red filter that transmits light between 620 and 700 mu for reading the transmission.

**Calculation:** Let U be density of unknown.
S be the density of standard (containing 0.04 mg P)
B be the density or reagent blank.

$$\text{mg of P per ml. of sample} = \frac{U-B}{S-B} \times 0.04 \frac{100}{1} \times 2$$

In animals normal phosphorus content is 4 to 8 mg per 100 ml. of serum. The indications for P estimation are the same conditions as enumerated for calcium.

**Hyperphosphatemia May be Encountered in the Following Condition.**

1. Severe renal disease.
2. Hypoparathyroidism
3. Excessive Vit. D feeding.
4. Terminal stages of parturient paresis.
5. EDTA poisoning.
6. A slight rise during healing of fractures.
7. In severe dehydration.

**Hypophosphatemia may be seen in**

1. Inadequate feeding of phosphorus.
2. Rickets
3. In inadequate feeding with Vit. D
4. Hyperparathyroidism.
5. During glucose utilisation.
6. After administration of insulin or adrenaline.

## MAGNESIUM
### (Method of Orange and Rhein Modified)

**Principle:** Magnesium is precipitated as the hydroxide in the presence of dye Titon yellow. The red colour formed is then compared with standards that have been treated in like manner. At an alkaline pH magnesium hydroxide reacts with titon yellow to form a reddish colloid. The intensity of the color is directly proportional to the coincentration of magnesium and follows Beer's law upto conentration of about 10 mg/100 ml polyvenyl alcohol is added as a stabilizing agent.

**Note.** Heavy metals interfere with color development and therefore glassware which has been used for such solutions like 'Nesslers' should never be used. Hemolysis should be avoided as there is a significantly greater concetration of magnesium in red cells than in serum. If the analysis is deferred or delayed the serum should always be separated from the red cells.

**Reagents Required**

1. **10% Trichloroacetic acid:** Dissolve 10 gms of trichloroacetic acid, reagent grade in 100 ml of water.
2. **Sodium hydroxide: 16%** (4 N): Prepare by diluting 16 gms of sodium hydroxide in 100 ml of ion-free distilled water.
3. **Titon yellow stock 0.35% :** Dissolve 350 mg of titon yellow in 100 ml of 50% isoproyl alcohol. Keep in polyethalene bottle in refrigerator.
4. **Titon yellow working 0.035%** : Dilute 10 ml stock to 100 ml with distilled water. Keep in polyethelene bottle in regregerator and prepare fresh every two weeks.
5. **Polyvenyl alcohol 0.1%** ; Dissolve 1 gm of polyvenyl alchohol in distilled water with the aid of gentle heating and stirring, cool and dilute it to 100 ml with water. Solution must be crystal clear.
6. **Magnesium standard stock** (1 ml = 0.5 mg ) Dissolve 0.5065 gms magnesium sulphate $7H_2O$) in distilled water and dilute to 100 ml.
7. **Magnesium standard working:** (1 ml = 0.015 mg)
   Dilute 3 ml stock standard to 100 ml with distilled water

**Procedure:**

(a) Unknown : Pipette 2 ml of serum into 15 ml centrifuge tube.

(b) Add 4 ml distilled water.

(c) Add 2 ml 10% trichloro acetic acid and mix. Centrifuge for 10 minutes at 1500 rpm. or filter through Whatman No 40 filter paper.

(d) Pipette 2 ml supernatant to a test tube.

(e) Add 1 ml distilled water.

(f) Standard: Pipette 1 ml standard, 1.5 ml of distilled water and 0.5 ml trichloro acetic acid into a tube.

(g) Blank: Pipette 2.5 ml distilled water and 0.5 ml trichloro acetic acid into a tube.

(h) Add 1 ml polyvenyl alcohol to each tube (i.e. to unknown, standard and blank) and mix.

(i) Add 1 ml Titon yellow working solution to each tube and mix.

(j) Add 1 ml sodium hydroxide (16%) to each tube and mix.

(k) Color develops immediately and is stable for 3 hours. Read at wave lenth of 560 m$\mu$. Adjust readings of blank at 100% transmittance of zero optical density.

**Caculations**

$$\frac{\text{Density of unknown}}{\text{Density of standard}} \times 3 = \text{mg. Magnesium per 100 ml of unknown}$$

Normal serum magnesium content in animals varies from 2 to 4 mg per 100 ml.

**Hypermagnesemia:** Mild elevated values may be noticed during parturition in old cows. In man slight increased levels were noticed in chronic nephritis and in oxalate piosoning.

**Hypomagnesemia** occurs due to inadequate consumption. In adult cattle and sheep it is found when grazed on lush pastures (which are deficient in Mg.) In calves that are wholly fed on milk for a long time hypomagnesemia may develop.

## SERUM ALKALINE PHOSPHATASE

**Reagents Required**

**Alkaline phosphate substrate:** In 100 ml volumetric flask introduce 3 ml of petroleum ether (B.P. 20 to 40$^0$), about 80 ml of distilled water, 0.5 gm of sodium Beta-glycero-phosphate, 0.424 gm of sodium diethyl barbiturate and water to the mark. The interface between the petroleum ether and the aqueous solution must be read while making up the volums to the mark. Put into glass stoppered bottle and layer an inch of petroleum ether on top and keep in a refrigerator.

**30 percent trichloroacetic acid:** Dissolve 30 gm of reagent grade trichloroacetic acid in water and dilute to 100 ml.

**5 percent trichloroacetic acid:** Dilute one volume of 30 percent solution with 5 volumes of distilled water.

**Standard phosphate solution:** Place 6.25 ml of the standard phosphate solution used for the determination of inorganic phosphate containing 0.5 mg of phosphorus in a 100 ml volumetric flask. Add 16.7 ml of 30 percent trichloroacetic acid solution and dilute to the mark with distilled water. This solution contains 0.04 mg of phosphorus in 8 ml of 5 percent trichloroacetic acid. When stored in cold, stable indefinitely.

Molybdate II and aminonaphtholsulphonic acid reagents same as for inorganic phosphorus estimation.

**Procedure: A. Incubated Sample**

1. Measure 9 ml of the alkaline phosphate substrate in a glass-stoppered cylinder or test tube and place in water bath at 37°C, until the temperature of the fluid in the test tube reaches the temperature of the water bath.
2. Add 1 ml of unhemolysed serum, mix and note the time of replacing the tube into the water bath.

3. Keep in the water bath exactly for one hour.

4. At the end of this time, remove the tube and keep for several minutes in ice water, add 2 ml of 30 percent trichloroacetic acid, mix well and let stand for a few minutes and then filter paper.

**B. Control Sample**

1. Near the end of incubation period in the above procedure measure 9 ml. of substrate into a glass stoppered cylinder and add 2ml of 30 percent trichloroacetic acid. Mix and add 1 ml of serum, mix , stopper, shake and filter as above. Transfer 8 ml of the two filtrates to 2 separate cylinders graduated at 10 ml In a third similar cylinder place 8 ml of standard phosphate solution, containing 0.04 mg of phosphorus. For photoelectric colorimetry, a blank is necessary. Thist is prepared by measuring in a similar cylinder 8 ml of 5 percent trichloroacetic acid. To each of the above tubes add 1 ml of molybdate II reagent and mix. Add later 0.4 ml of aminonaphtholsulphonic acid reagent, dilute immediately to the 10 ml mark with distilled water, mix and allow to stand for 5 minutes for the development of the color.

**Calculation**

**Visual colorimeter:** Both the control and the unknown should be compared with the standard.

$$\text{mg. of inorganic P in 100 ml. of serum (control or incubated)} \quad \frac{\text{Reading of standard}}{\text{Reading of unknown}} \times 0.04 \times \frac{3}{2} \times 100$$

**For photoelectric colorimetry**

$$\text{mg. of inorganic P in 100 ml. of serum (control or incubated)} \quad \frac{\text{Density of unknown}}{\text{Density of standard}} \times 0.04 \times \frac{3}{2} \times 100$$

The Phosphatase activity is the difference between the inorganic phosphate content of the incubated serum and the control sample, expressed in mg. ofphosphorus per 100 ml. So is the control serum has 3 mg. percent and the incubated sample 8 mg percent, the phosphatase activity is 8 minus 3 or 5 Bodansky units per 100 ml of serum.

The serum alkaline phosphatase in normal cattle and sheep is very wide and so is not of practical utility. In dogs it is 3 to 6 Bodansky units and so phosphatase activity can be utilised to assess the health of the liver.

Normally, serum alkaline phosphatase owes its origin to the osteoblasts. Some of it is normally excreted in the bile. So if there is interference of bile flow, the serum alkaline phosphatase activity is said to be increased. In some abnormal conditions, for example when there is biliary stasis, excessive alkaline phosphase or one of its activators may be formed by the liver.

**Increased values are seen in the following conditions:**

1. Billary obstruction.

2. Liver diseases — hepatic necrosis and CCI administration.
3. Bone diseases:
   (a) Rickets
   (b) Osteomalacia
   (c) Hyperparathyroidism
   (d) Osteogenic sarcoma.

For the dog, serum alkaline phosphatase activity can be used as a liver function test.

## Serum Glutamic Pyruvic Trnsaminase and Serum Glautamic Oxaloacetic Transaminase. (SGPT & SGOT)

GOT is distributed widely in many tissues but liver , myocardium and skeletal muscles are rich in this enzyme. When there is damage to the cells of these tissue, the enzyme, escapes into the blood and so the SGOT enzymatic activity increases. Glutamic-pyruvic transaminase is present in high concentrations in the liver of cat and dog but little is present in livers of horse, cattle and pigs. So in liver necrosis of dogs and cats the GPT escapes into the blood and so the SGPT activity increases.

Advantage of this is taken to assess the health of the liver in dogs and so this is one of the liver—function testsd in this animal.

### Principle

When SGOT is added to a mixture of alpha-ketoglutaric acid and aspartic acid, the enzyme catalyses the transfer of alpha-amino group of aspartic acid and the alpha-ketoglutaric acid and thus oxaloacetic and glutamic acids are formed.

Similarly, SGPT catalyses the transfer of alpha-amino group of alanine and alpha-ketoglutraic acid when added to a mixture or these two amino acids, pyruvic acid and glutamic acid are formed.

### Reagents Required

**0.1 M phosphate buffer:** Dissolve in a liter volumetric flask 13.97 gms of $K_2HPO_4$ and 2.098 gm $NAH_2PO_4$ and make upto the mark with distilled water. This has a pH of 7.4.

**Stock alpha-ketoglutaric acid solution:** In a 100 ml volumetric flask, dissolve 0.292 gm of alpha-ketoglutaric acid in 20 ml of distilled water, adjust the pH to 7.4 with 1 N NaOH and then make upto the mark with distilled water.

**1 N NaOH (Approximate):** Dissolve 41 gm of NaOH in a litre of distilled water.

**Buffered aspartic acid solution:** Dissolve 2.955 gm of DL aspartic acid and in 20 ml of 1 N NaOH, adjust pH to 7.4 and make up the volume to 100 ml with 0.1 M phospgate buffer.

**GOT substrate** Dilute 1 volume of stock alpha-ketoglutaric acid solution with 9 volumes of buffered aspartic acid aspiartic acid solution just before use.

**Sodium pyruvcate solution** Dissolve 22 mg of sodium pyruvate in 100 ml of 0.1 M phosphate buffer prepare this before use.

**2-4-Dinitrophenylhydrazine solution** Dissolve 19.8 mg of reagent grade 2,4-diniro-phenylhydraziune in 10 ml of concentrated MCI and then dilute it to 100 ml with distilled water. Store in amber colored bottle.

**0.4 N NaOH:** Dissolve 1.977 gm of DL-alanine in 20 ml of distilled water. Adjust the pH to 7.4 with 1 N NaOH and ditute to 100 ml with 0.1 M phosphate buffer.

**GPT substrate:** Dilute I volume stock alpha-ketoglutaric acid with-9 volumes of buffered alanine. Prepare just before use.

## SGOT Activity

### Procedure

1. In to a test tube pipette 1 ml. of SGOT substrate and keep in a water bath at 37°C.
2. After this has attained the temperature of the water bath add 0.2 ml of serum. Mix by gentle rotation and keep in the water bath for exactly 1 hour.
3. Remove from water bath and add 1 ml of 2,4-dinitrophenlhydrazine solution. (color reagent)
4. **Control** Towards the ends of incubation period in step 2 prepare a control with 1 ml of substrate plus 0.2 ml of serum and 1 ml of color reagent.
5. **Standard** To 0.6 ml. pf tjes substrate add 0.2 ml of distilled water and 0.4 ml of sodium pyruvate solution and 1 ml of color reagent.
6. **Blank** To 1 ml of substrate add 0.2 ml of distilled water and 1 ml of color reagent.
7. 20 minutes after the addition of the color reagent, add to all the four tubes 10 ml of 0.4 N NaOH and mix by inversion. Keep for 10 minutes and then read the colorimeter.
8. If it is Klett-Summerson, use a green filter after adjusting to zero with distilled water.

If it is Spectronic, 20, read at 505 mμ reducing the optical density to zero with distilled water blank.

### Calculation

$$\frac{\text{Reading of unknown} - \text{reading of control}}{\text{Reading of standard} - \text{reading of blank}} \times 40 = \text{units per ml}$$

## SGPT Activity

1. In a test tube measure 1 ml of GPT substrate and keep in a water bath at 37°C for 5 minutes (i.e. till the fluid attains the temperature of the water bath)
2. Add 0.2 ml of serum and mix gently and keep in a water bath for exactly 30 minutes.

3. Remove it from the water bath and then add 1 ml of the color reagent.

4. **Control :** To 1 ml of GPT substrate add 0.2 ml of the serum and 1 ml of color reagent at end of the incubation period of step 2.

5. **Standard:** To 0.6 ml of GPT substrate add 0.2 ml of distilled water 0.4 ml of sodium pyruvate solution and 1 ml of color reagent and mix.

6. **Blank.** To 0.2 ml of distilled water add 1 ml of the GPT substrate and 1 ml of the color reagent and mix.

7. 20 minutes after addition of the color reagent, to all the tubes add 10 ml of 0.4 N NaOH, mix by inversion and keep for 10 minutes.

8. At the end of this time read in the colorimeter as described for SGOT activity and calculate the units per ml of serum in the same way.

| Normal values | SGOT | SGPT | units (or ug pyruvate |
|---|---|---|---|
| Dog | $20 \pm 7$ | + | formed per ml of serum) |

SGOT activity. This is of value only in the dog.
Increased SGOT activity is seen in dog in the following conditions.

1. Hepatic necrosis
2. Myocardial infarction
3. Muscular dystrophy

**SGPT**

Elevation of SGPT activity is seen in liver nercrosis of the dog. Since heart and kidneys also contain GPT, one should conduct BSP clearance test also to confirjm hepatic damage. BSP clearance is delayed in hepatic necrosis and fibrosis while SGPT is elevated only in hepatic necrosis. Elevation of SGPT has been found in the following conditions in dogs: intrahepatic cholestasis: complicated Babesia canis infection; hepatic neoplasia, fatty degeneration of the liver; leptospirosis: carbon tetrachloride poisoning; suppurative hepatic necrosis, arsenic poinsoning; hepatic necrosis seconday to pyometra, and extensive hepatic malignancies.

As per Cornelius, the normal values in dogs are 10-53 units.

In moderate hepatic necrosis 50-400 units.

In servere necrosis of liver above 400 units.

## SERUM CHOLESTEROL
(Method of Zak)

**Principle**

Method of Zak depends on the principle of Liberman-Buchard reaction in which an acetic acid solution certain sterols produce red color when treated with Ferric chloride and sulphuric acid. The color produced is estimated colorimetrically.

**Procedure**

Add 0.1 ml of serum to 6 ml of Ferric chloride precipitating reagent and gently shake. Filter through No 42 filter paper. Prepare final standard by mixing 2 ml of cholesterol working standard (0.1 mg per ml) and 4 ml of Ferric chloride blank solution. To 3 ml each of filtrate, final standard and Ferric chloride blank solution add 2 ml concentrated sulphuric acid slowly. A ring is formed at the interphase. Shake gently. Cool. Compare at 470 mu.

**Result :**

$$\frac{\text{Reading of unknown}}{\text{Reading of standard}} \quad \frac{\text{Concentration of standard}}{\text{Volume of unknown}} \times \frac{100}{1} = \text{mg. of cholesterol}$$

**Regeants**

1. **Stock Ferric chloride colution:** 840 mgs of Ferric chloride is dissolved in a few ml of glacial acetic acid and diluted to 100 ml with acetic acid.

2. **Ferric chloride precipitating reagent:** Stock Ferric chloride solution is diluted 1 in 10 with glacial acetic acid.

3. **Ferric chloride blank:** 1.7 ml of stock Ferric Chloride diluted to 20 ml with glacial acetic acid.

4. **Cholesterol stock standard:** 100 mg pure dry cholesterol is dissolved in 100 ml of glacial acetic acid.

5. **Working standard:** 2 ml of the cholesterol standard mixed with 1.7 ml of stock Ferric choride solution and diluted to 20 ml with glacial acetic acid.

6. **Final standard:** This is prepared by mixing 2 ml of working standard and 4 ml of Ferric chloride solution — just before use.

Cholesterol exists as free cholesterol and cholesterol esters when combined with fatty acids. In the blood serum, total cholesterol ranges, among animals from 50 to 250 mg per 100 ml.

It is the liver that is responsible for the maintenance of homeostasis of serum cholesterol. Cholesterol estimation is indicated in suspected hepatic and thyroid dysfunction.

**Hypercholesteremia or Hypercholesterolemia**

Increased blood cholesterol values are found in the following conditions. Chronic and acute nephritis, chronic nephrosis, Diabetes mellitus, hjigh lipiud intake, pregnancy, cholelithiasis, hypothyroidism, biliary obstruction and corti-sone therapy.

**Hypocholesteremia**

Decreased cholesterol values are met with in the following conditions, hyperthyroidism, low fat diet, starvation, acute infections, liver diseases and anemia.

NORMAL DATA

| | Incubation period Days | Age of maturity months | Number of eggs laid (Yearly) |
|---|---|---|---|
| Fowl | 21 | 6 | 150-200 |
| Duck | 28 | 7 | 80-300 |
| Goose | 30 | 10-11 | 24-50 |
| Turkey | 28 | 10-11 | 25-70 |
| Gunia-fowl | 28 | 9 | 50 |
| Pigeon | 18 | 7 | 2 in a brood: 2-3 brood in a year, |
| Pea-fowl | 18 | 12 | 15 |
| Pheasant | 25 | 10-12 | 8-15 |
| Ostrich | 35-40 | 48 | 16-18 (2 broods yearly) |

# 3

# URINE

---

| | |
|---|---|
| Color | Ketone bodies |
| Transparency |     Rothera's test |
| Odor |     Ketonuria |
| Specific Gravity | Blood |
| Reaction |     Peroxidase test |
| Proteins |     Hematuria |
|     Robert's test |     Hemoglobninuria |
|     Hiller's test | Bile pigments |
|     Proteinuria |     Gmelin's test |
| Glucose | Sediments |
|     Benedict's test |     Microscopical examination |
|     Glycosuria | |

---

we have already learnt that kidneys play an important part in the regulation of water balance, acid -base balance, electrolyte balance and in the manitenance of osmotic presure of the body fluids.

When kidneys are damaged, their function is altrered, which is reflected by the alteration in the quantity and quality of the urine excreted. Hence examination of the urine helps in assessing the health of the kidneys. At least 4 ouncesof urine must be collected. It can be preserved in the refrigeratior. If this is not possible, a drop of toluene or a crystal of thymol must be added. If the structure of the cells and casts is to be mantained in the sediment, then add adrop of 40% formaline to one ounce of urine. If the sample of urine is small, it should be centrifuged first to ensure an adequate amount of sediment and the supernatent can be used for chemical tests. Casts and erythrocytes disappear upon standing and as examination should be made on fresh speciments. The following are the date examined for.

## COLOR

The color of the urine may be any one of the following: Colorless, pale yellow, yellow, dark yellow, yellowish-brown, greenish-yellow, red, reddish-brown, brown green, blue, milky. The yellow color is due to urochromes.

Pus in urine gives a thick color.

Dark yellow color is due to dehydration as occurs in vomition, diarrhoea, serous exudations, fever and in deficient intake of water.

Pale urine is usually seen in diabetes mellitus, diabetes insipidus, when too much of water is ingested, in pyometra (When there is increased water consumption) and in chronic interstitial nephritis in dogs.

In jaundice, when bile pigments are present, the urine is yellow, brown or greenish yelow.

Presence of hemoglobin and blood gives the urine a red color.

In congential pophyria the urine has a faint pink color. So also in phenothiazine medication, urine has a red color.

Greenish color of urine is found when animals are medicated with methylene blue or Diazan.

## TRANSPARENCY

The transparency of the urine is tested by viewing it against light in a test tube. Normally it should be transparent and clear. But in the horse, it is cloudy and thick due to the presence of mucus and calcium carbonate crystals. The horse urine, therefore, must first be allowed to settle. The supernatant urine is then used for other tests. ( After the mucus and crystals have settled).

The urine is cloudy if it contains:

1. Epithelial cells in large numbers.
2. Leucocytes in large numbers as in pyuria
3. Bacteria in large numbers. The cloudiness cannot be got rid of by filtration or on standing
4. Mucus as in the urine of the horse.
5. Crystals; calcium carbonate or amorphous urates or amorphous phosphates.
6. Blood, when present gives a red or brown smoky appearance.

**Note :** In a dog if the freshly voided urine is cloudy then it is pathological due to the presence of pus, blood and mucus.

## ODOR

The ordor of the urine is due to volatile acids. It becomes ammoniacal due to decomposition.

Sweetish ordor is got in diabetes mellitus. In ketonuria, there is odor of acetone. In turpentine medication, the urine may have odor of violets.

Foam : Normal urine may foam if shaken. But in cases where there is albuminuria, foam occurs in plenty.

## SPECIFIC GRAVITY

This is a measure of the relative amounts of solids in solution in the urine. So it indicates the capacity of the kidney to concentrate the urine by reabsorption.

Generally, sp. gr. varies inversely with the quantity of urine. Bxception is diabetes mellitus in which due to the presence of sugar, the sp.gr. is high though the quantity is also high.

In domesticated animals the following are the normal values.

| Animal | Range | Average |
|---|---|---|
| Horee | 1020 to 1050 | 1035 |
| Cattle | 1025 to 2045 | 1035 |
| Sheep | 1015 to 1045 | 1030 |
| Goats | 1015 to 1045 | 1030 |
| Dogs | 1015 to 1045 | 1025 |
| Cats | 1020 to 1045 | 1030 |
| Pigs | 1010 to 1030 | 1015 |

Specific gravity is determined by means of urinometer. This consists of a cylinder, and float on the stem of which is a scale starting from top to neck of the stem.

### Procedure

1. Fill the cylinder with urine upto about one inch from the top. Take care that there are no air bubbles since the presence of these at the top interferes with the reading.
2. Place the float in the urine and spin it so that it finds its level.
3. Read the mark on the stem that is in level with the highest level of the meniscus of the urine. This gives the sp.gr. of the urine.

To test the accuracy of the urninometer, use distilled water instead of urine in the above test. The reading must be 1000 at the room temperatre (20°C)

### Increase in Sp. gr. Occurs in the Following Conditions

Usually increase in solids in urine raises its sp. gr.

1. In acute interstitial nephritis, the sp. gr. is elevated dur to the inability of the kidney to excrete water (sp. gr. varies from 1030 to 1060).
2. In cystitis due to the presence of proucts of inflammation.
3. Diabetes mellitus—due to glycosuria,
4. Dehydration vomition, diiiiiarrhoea, fever, reduced fluid intake.

**Decreased Sp. Gravity**

1. In increased fluid intake—this may be physiological.
2. Inability of the kidney to concentrate: Chronic interstitial nephritis in dogs, sp. gr. is 1010 or less.
3. Advanced stages of uremia.
4. Diabetes insipidus.

## REACTION OF THE URINE

The reaction of the urine is dependent on the diet. In vegetarians the reaction is alkaline. With carnivores and omnivores it is acidic. The normal reaction of urine among animals:

Horse — alkaline (pH 8)
Cattle — alkaline (pH 7.4 to 8.4)
Sheep — alkaline
Pig — acid or alkaline, depending upon the feed.
Dog — acid (pH 6 to 7)
Cat — acid (pH 6 to 7)

Reaction is detected using a litmus paper.

Acid urine turns blue litmus red.
Alkaline urine turns red litmus blue.

**Acid urine is encountered in the following situations:**

1. Normally in carnivore urine.
2. In starvation.
3. Fever.
4. In the young calves and foals that are on milk.
5. Acidosis-metabolic and respiratory.
6. Prolonged muscular activity.
7. Administration of acid salts: ammoniam chloride, sodium chloride, calcium chloride. Sodium acid phosphate.
8. Diet with excess of protein.
9. In eachectic conditions.

**Alkaline urine is seen in the following situations:**

1. In herbivores it is normal.

2. Cystitis.
3. Retention of urine.
4. Administration of alkaline salts: acetate, bicarbonate, citrate or nitrite of sodium or potassium.
5. Rapid absorption of transudates.

***Note:*** In urine which is alkaline, ammonia is less readily soluble and hence ammoniacal smell in a freshly voided urine in dog is diagnositic of cystitis and in such cases pH will be high.

## PROTEINS

Normal urine dose not contain any protein.

**Tests for detection of albumin in urine.**

**1. Robert's Test**

Robert's reagent: Concentrated nitric acid 1 part, saturated magnesium sulphate 5 parts (prepared by dissolving 770 gm of mag sulph to one litre of water).

(i) Take 2 ml of the reagent in a test tube.

(ii) With a pipette layer carefully on the reagent, by allowing the urine to flow down the sides of the tube, held slantingly, 2 ml of the urine. (Note—if the urine is cloudy centrifuge or filter before you do the test).

(iii) At the juction of the two fluids a white ring forms if albumin is present.

**2. Heller's Test**

This is very commonly used (though not so sensitive as the Robert's)

(i) Take 2 ml of concentrated nitric acid in a test tube.

(ii) With a dropper or a pipette layer on top, 2 ml of urine.

(iii) At the junctionof the two liquids a white ring is formed if urine contains albumin.

**Proteinuria may be present in the following conditions**

(a) **Physiological:** is a transient episode due probably to the increased glomerular filtration consequent on congestion of capillaries:

seen in

1. Excessive muscular exertion.
2. Excessive ingestion of proteins.
3. In times of emotional stress.
4. In convulsion.

(b) **Pathological:** In nepritis, because of increased permeability of the glomerular filter.

Acute interstitial nephritis —There is protein in the urine with casts.

Chronic interstitial nephritis—Slight proteinuria with casts.

Pyelonephritis: There is marked proteinuria with leucocytes and red blood cells.

Nedphrosis: *Due to poisoning* by Phenol, arsenic, lead, phosphorus, mercury, sulphonamide, turpentine, ether, bismuth, salicylic acid—proteinuria is excessive.

**In renal congestiom due to:** cardiac congestive failure: pressure on abdominal veins due to ascites or tumors, there is slight proteinuria.

In amyloidosis, renal infraction and neoplasms also there is proteinuria.

**Post renal conditions in which proteinuria is present :**

In the following conditions protein gains entrance after it has left the kidney:

## GLUCOSE

Normally urine does not contain any glucose. All of it that is found in the glomerular filtrate is reabsorbed by the tubular epithelium since glucose is a thereshold substance.

Glucose in urine is called glycosuria. It is detected by:

**Benedict's Test**

Benedict's qualitative reagent is supplied by the Medical Stores Depot. This contains copper sulphate. Glucose is a reducing agent, which converts the cupric ions in the copper sulphate to cuprous ions-cuprous oxide.

1. Take 5 ml of the reagent in a test tube and heat it. If it has not changed color (usually it should not).
2. Add 8 drops of the urine to it and boil it carefully, seeing that it does not spurt while boiling. Then if glucose is present, a greenish or yellow precipitate forms.

**Glycosuria is present in the following conditions**

1. In emotional states—fear, excitment—in these conditions, there is sudden release of adrenaline which causes hyperglycemia with resultant glycosuria.
2. Heavy meal of carbohydrates.
3. Diabets mellitus.
4. After general anesthesia.
5. Hyperthyroidism—due to rapid absorption of glucose from the bowel.
6. Chronic pancreatic necrosis.
7. Acute pancreatic necrosis.
8. Hyperpituitarism—produces hyperglycemia.
9. Overactivity of adrenal cortex.
10. Shock.
11. Administration of glucose intravenously.
12. Chronic liver disease.

13. Enterotoxomia in sheep.
14. Brain tumors affecting floor of fourth ventricle.

False positive reaction for glucose can occur if other reducing agents are present; antibiotics, lactose, ascorbic acid, salicylates, chloral hydras, morphine, formaldehyde, uric acid.

## KETONE BODIES IN URINE

These are acetoacetic acid ($CH_3COCH_2COOH$)
Beta hydroxy butyric acid ($CH_3CHOHCH_2COOH$)
Acetone ($CH_3COOH_3$)

These are not normally present in the urine. But when fat metabolism is impaired (due to deficient carbohydrates), ketone bodies accumulate in the body giving rise to ketonemia and then ketonuria results.

Ketones in the urine are detected by Rothera's test.

Saturate 5 ml of urine with ammonium sulphate. Add 2 or 3 drops of 5% sodium nitroprusside solution and then add strong ammonium hydroxide. A permanganate color is indicative of ketones.

**Ketonuria is present in**

1. Diabetes mellitus---in dogs and cats.
2. High fever.
3. In starvation.
4. In cows it is met with in acetonemia, milk fever and in anorexia of high yielding animals.
5. In ewes in pregnancy disease.
6. Cachectic conditions.

## BLOOD

Hemoglobin contains a peroxidase which can liberate oxygen from hydrogen peroxide. This oxygen gives a blue color in the presence of benzidine.

**Test**

1. In a test tube dissolve a small quantity of benzidine in glacial acetic acid.
2. Add 2 ml of urine to this.
3. Add one ml of fresh hydrogen peroxide and mix.

If blood is present a blue or green color develops.

**Hematuria** is the presence of whole blood in urine.

**Hemoglobinuria** is the presence of hemoglobin in urine. These can be differentiated by centrifuging a specimen of urine. In hematuria, the red blood cells will be layered at the bottom while the supernatant liquid is colorless. In hemoglobinuria in which the erythrocytes are laked, the whole fluid will be red.

**Hematuria is encountered in the following conditions**

1. Pyelonephritis.
2. Acute nephritis.
3. Ureteritis.
4. Cystitis.
5. Pyelitis.
6. Urolithiasis.
7. Passive congestion of kidney.
8. Infection of kidney.
9. Abscess of kidney.
10. Neoplasms of kidney, bladder or prostate.
11. During estrus.
12. At post partum in females.
13. Prostatitis.
14. Trauma to urethra-duringg improper cartheterisation.
15. Severe infections as in anthrax, leptospirosis, Rubarth's disease.
16. Toxic chemical agents: copper poisoning, Phenol, sulphonamide, mercury poisoning, arsenic poisoning thallium poisoning.
17. Sweet clover disease.
18. Shock-capillary hemorrahage.
19. Parasites—Dioctophyma renale in dogs.

If blood is found in the last drops of urine then the source is bladder.
If the urine is red throughout, the source of blood is kidney.
If the first portion of urine is red, then the source of blood is some urethral lesions.

**Hemoglobinuria**: In this condition, there is increased hemolysis and is found in the following conditions.

1. Leptospirosis.
2. Post parturient hemoglobinuria.
3. Babesiosis.
4. Photosensitisation.
5. Bacillary hemoglobinuria — Clostridium hemolyticum infection.
6. Chemical hemolytic agents: Copper, mercury, sulphonamides.
7. Severe burns.
8. Hemolytic disease of the neew born.
9. Incompatible blood tranfusion.
10. Drinking large volume of water.
11. Plant poisons: Ranunculus, ash, hazal, hellebore etc.

In myoglobinuria (azotorial) when the urine contains myoglobin, a positive test is obtained with the benzidine test. This can be differentiated from hemoglobinuria by the abscence of erythrocytes in the sediment.

## BILE PIGMENT

### Gmelin's Test

1. In a test tube take 2 ml of concentrated nitric acid.
2. Carefully on top layer 2 ml, of urine with a pipette.
3. At the junction of the two fluids a play of colors—green, violet etc. will be observed if bile pigment is present.

## UROBILINOGEN

Normal urine contains urobilinogen. But in obstructive jaundice, urobilinogen, is not present in the urine.

### Test for urobilinogen

**Reagent required** Ehrlich's aldehyde reagent; Dissolve 2 gms of paradimethylaminobenzaldehyde in 100 ml of a 20% aqueous solution of concentrated hydrochloric acid.

To 5 ml of the urine in a test tube add 0.5 ml of the Ehrlich's aldehyde reagent, A pink color is positive for urobilinogen.

## MOCROSCOPICAL EXAMINATION OF URINARY SEDIMENT

Always urine examination should include the examination of the sediment. Since the casts and erythrocytes disappear if kept standing, the sediment test should be done immediately after collection of the urine.

The following are seen in the sediment, which is obtained by centrifuging a sample of the urine; epithelial cells, erythrocytes, leucocytes, casts, microorganisms, mucous threads, crystals, spermatozoa and fat.

### Epithelial Cells

Squamous epithelial cells: These are the largest and are derived from the superficial layer of the mucosa of the urethra and vagina and occur singly or in sheets, and contain a small round nucleus.

Transitional epithelial cells are smaller than the epithelial cells but larger than the renal cells; have round spindle shape or caudate form; have a granular appearance and a small round nucleus; and are derived form the mucosa of the bladder, ureters and pelvis of the kidneys.

Renal epithelial cells are small and round with a single nucleus. These are not clearly defined.

Normally it is usual to find some epithelial cells and the transitional epithelial cells, especially in the female urine.

Pathologically, large number of renal epithelial cells are found in acute interstitial nephritis. Transitional epithelial cells are found in cystitis, and pyelenephritis.

**Erythrocytes:** Presence of these indicates hemorrhage in some part of the genitourinary tract. In faulty catheterization, erythrocytes may be found in urine.

**Leucocytes:** In normal urine only a few are present. They are larger than the erythrocytes and smaller than the epithelial cells. They have granular cytoplasm.

Increased number of leucocytes in the urine is known as pyuria. If there are 10 or more of leucocytes per a high power field, it is indicative of pyuria. Pyuria is found in nephritis, pyelonephritis, pyelitis, ureteritis, cysitis, urethritis, or contamination from vulvitis, vaginitis or balanitis.

**Casts:** We have already described the casts (page 399 of Pathology book). Presence of casts indicates renal injury. *Casts are rarely seen in the alkaline urine of herbivores.*

**Mucous threads:** In the urine of horses, there are long mucous threads normally. In animals presence of mucous threads is indication of contamination from the genital secretions.

**Parasites:** The eggs of parasites that may be seen in the urine are those of :

1. **Stephanurus dentatus,** which is in swine kidney.
2. **Dioctophyma renale,** which is in the dog kidney.
3. **Cappillaria plica,** the bladder worm of dog, cat & fox.
4. If the urine is contaminated with feces, a variety of eggs will be found.

**Crystals**

In acid urine are found amorphous urates, uric acid, calcium oxalate and hippuric acid.

In alkaline urine are found triple phosphates, amorphous phosphates, calcium carbonate and rarely ammonium urate.

In acute liver disease leucine and tyrosine crystals may be seen.

In the Dalmatian dog, uric acid and urate crystals are seen.

**Bacteria:** Urine may contain large number of bacteria in cystitis, pyelonephritis and genital tract infections.

Yeasts and fungi are only contaminants.

## MASTITIS TEST

**Fat:** Is seen as round and highly refractile globules. These can be stained with Sudan III, when they take orange stain.

Fat in urine is found in:

1. Diabetes mellitus
2. High fat conten of diet.
3. Obesity.
4. Hypothyroidism
5. Contamination from extraneous sources.

# 4

# MASTITIS TESTS

| | |
|---|---|
| Collection of sample | Whiteside test |
| Strip cup test | Chloride test |
| Bromthymol Blue test | Hotis test |
| California Mastitis test | Antibiotic Sensitivity test. |

In mastitis the quality (and quantity) of milk is altered. Diagnosis of mastitis is facilitated by means of certain simple tests that can be performed easily in the field.

Colostrum and milk for a week after parturition are not suitable for tests to be conducted since they do not represent normal milk and contain abnormal quantities of protein and may react to tests for mastitis.

Normal milk does not contain any visible particles. The normal pH of the milk is 6.4 to 6.8. Milk is isotonic with blood plasma. In mastitis and during late lactation period and dry period, the concentration of lactose and casein is reduced and sodium chloride and sodium bicarbonate pass into the alveoli from plasma. So in these situations, the milk becomes more alkaline and contains greater amounts of chlorides. (The normal chloride content of milk is 0.08 to 0.15 gms%.)

## Collection of a Sample of Milk

1. Clean the under well with warm water.
2. Wipe it with a clean cloth soaked in disinfectant.
3. Dry.
4. Paint the test orifice with tincture of iodine or 70% alcohol and allow to dry.
5. Label sterile tubes or bottles RF, RH, LH, LF.

6. Collect direct from the test into the bottle, stopper it and use the milk immediately for the tests. If tests are delayed, keep in a refrigerator. At least 5 ml of milk is required.

## PHYSICAL TESTS

Mastitis milk will have clots and may also be colored.

**Stripcup test:** Strip cup is the most useful thing in a dairy farm. It is a good practice to use it daily at every milking. The advantages of doing so are : that early cases of mastitis can be diagnosed, that the fore milk which contains bacteria can be discarded.

The strip cup consists of a flat enamel plate, partitioned into 4 areas. This plate has a black color, so that the clots amy be easily seen. While milking, milk the first 2 or 3 streams from the teat into this cup. Have always a regular system. Starting from left Fore teat, then pass on to the Left Hind, then Right Hind and lastly Right Fore. If clots are seen, remove the animal from the herd immediately, isolate and then investigate more thoroughly later.

### Bromthymol Blue Test

In mastitis, the reaction of milk changes to alkaline pH, i.e. it changes from normal (6.4 to 6.8) to a pH of nearly 7.4. This can be detected by Bromthymol Blue.

A. Prepare Bromthymol Blue solution as follows:
   (i) Dissolve 1.0 Gms of Bromthymol Blue in 160 ml of cf N/100 NaOH.
   (ii) Make up solution with 590 ml distilled water.

B. Test milk:
   1. Take graduated 10 ml cylinder or mark in a test tube at 6 ml capacity.
   2. Take 1 ml of the Bromthymol Blue solution, in this tube.
   3. Add to this 5 ml of freshly drwan milk. (Milk kept for sometime is of no use).
   4. Mix by inversion once or twice.
   5. If milk is normal the color will be yellow or slightly yellowish-green. But if the color is green or dark green or blue, then the milk contains inflammatory exudate and so mastitis is present.

**Note:** Cows in the later stages of lactation may give false positive test.
Some companys market Bromthymol blue papers for ready and easy use.

## CHEMICAL TESTS

### California Mastitis Test

This test is based on the increassed leucocyte counts during mastitis as well as to the increased alkalinity obtaining in mastitis. These altered characteristics are due to the inflammatory exudate that forms in the alveoli and which gets mixed with the milk.

The exudate contains basic salts and hence the reaction turns alkaline.

**Procedure:** This test requires a plastic paddle with four chambers. The testing reagent is dispensed in plastic bottles provided with a fine nozzle. On squeezing the bottles a jet of the reagent can be addd into the chamber.

1. Prepare the under as mentioned above.
2. Into the recepticles of the plastic paddle, milk directly form the tests.
3. Squirt from the plastic bottle equal quantity (with the milk) of the reagent, and then by rotatory movement of the paddle, mix the milk with the reagent. Reactions occur immediately. In positive cases, the color turns greenish blue in alkaline ranges. Due to increased leucocyutes, there may be a preciptate or a gel formation.

The original reagent formulated by Schalm and Noorlander contained alkylaryl-sulphates and sulphonates of sodium and potassium which are difficult to be obtained. So, these where substituted by the detergent "Teepol" which gave equally good results. It is made as follows.

| | | |
|---|---|---|
| Sodium hydroxide | ... | 1.5 gms |
| Teepol (Shell chemical) | ... | 0.5 ml |
| Brom thymol blue | ... | 0.01 gm |
| Distilled water to make | ... | 100 ml |

Teepol is easily available and so it is easy to prepare this reagent.

Glaxo Laboratories are marketing a product, "Mastaid" which can be used for this test.

**White side test:** This can also be performed easily, and the test is dependent upon increased leucocyte content of mastitis milk.

The following are required;

1. A glass plate with under side painted black.
2. 2 droppers which can deliver equal drops.
3. A broom stick.
4. 4% sodium hydroxide solution. (Keep in a pyrex or corning glass to maintain the exact concentration).

**Procedure**

1. Place 5 drops of suspected milk on the glass plate.
2. Add 2 drops of 4% sodium hydroxide reagent.
3. Rapidly stirr with the broom stick for 20 to 25 seconds.

In chronic cases, white flecks are found. But in normal milk, there will be only a milky, opaque fluid.

**Chloride test:** This is another test that can be easily performed. This test depends upon the increased quantity of chlorides that mastitis milk contains. Normal chloride content of milk is 0.08 to 15 gms%. But in mastitis due to the presence of inflammatory exudate, the chloride content will be higher.

**Prepare** Solution. A. Silver nitrate 1.3415 gms
Distilled water 1000 ml
Keep in amber colored bottle

Solution. B. Potassium chromate 10 gms
Distilled water 100 ml

**Procedure**

1. To 1 ml of milk in a test tube add 5 ml of solution A.
2. And 2 drops of solution B to this mixture and mix.

If color is yellow , the milk contains more than 0.15 gms% of chlorides. (**Note:** False positive reaction may be obtained in milk of cows in early or late lactation).

## HOTIS TEST

With this test it is possible to detect the presence of *Streptococcus aglactiae*, which ferments lactose of milk yielding acid. So the indicator used. Bromeresol purple. is changed to yellow.

1. In a test tube keep 0.5 ml of sterile bromcresol purple solution (0.9 % in distilled water.)
2. Add 9.5 ml. of fresh milk.
3. Incubate at 37°C for 24 hours.
4. If the sample of milk contains **Str. agalactiae**, there will be canary yellow colonies of the orgainsms along the sides of the tube.

## ANTIBIOTIC SENSITIVITY TEST

For effective treatment of mastitis, it is imperative to select an antibiotic to which the causative organisms are sensitive, i.e. an antibiotic that can destroy all the organisms. For this antibiotic sensitivity test can be performed, with which we can know, within 4 hours, which antibiotic is most efficacious.

1. Mix well one ml of suspected milk with melted brain-heart infusion agar.
2. Pour the misture on 20% blood agar plates.
3. After the agar has hardened, place the antibiotic discs with a sterile forceps an the agar, pressing them lightly into position.
4. Incubate at 37°C for 4 hours and then read the result.
5. Effective antibiotic will not allow the growth of the microoraganisms and so the hemoglobin is intact ( not hemolysed). So, the area around such effective antibiotic discs will be bright red, while the area around the ineffective discs will be dark red as the hemoglobin is reduced by the action of the bacteria, which proliferate without any hindrance.
6. Continue incubation for 24 hours. Now clear zones of inhibiton of growth of

# 5

# LIVER AND KIDNEY FUNCTION TESTS

bacteria will be noticed around effective antibiotic discs

| | |
|---|---|
| Iodine test | van den Bergh test |
| Takata—Ara test | Bromsulphalein Dye excretion test |
| Clotting time, estimation of | BUN estimation of |

The following tacts should be borne in mind while deciding to conduct liver function tests.

1. In diseases of liver, the several functions of the liver are not equally affected. So it may be necessary to do several tests to measure different functions.

2. The liver has great reserve powers. So unless 80% of the liver is damaged. there may not be any evidence of hepatic dysfunction as measured by the tests.

3. It has been stressed, again and again, that the hepatic tissue has great regenerative capacity and so hepatic tissue destroyed is quickly replaced.

Liver function tests require very good laboratory facilities and all cannot be easily performed under the conditions of the field. The tests that can be easily performed are described here.

## THE IODINE TEST

| | | |
|---|---|---|
| The reagent. | Iodum (crystals) | 1 gm |
| | Potassium iodide | 2 gm |
| | Distilled water | 27 ml |

Keep in an amber colored bottle.

**Procedure**

1. Place a drop of the serum on a glass plate.
2. Add a drop of the reagent and stir with a broom stick.

In positive cases flocculi occur at once. Depending upon the intensity of flocculation, it may be graded as 1 plus, two plus or three plus. A one plus flocculation consists of small flocculi, A three plus reaction has the whole mixture curdled.

This test has been found to indicate the existence of cirrhosis in cattle (aflatosicosis) and hepatic damage.

## TAKATA-ARA TEST

1. In a rack, keep 8 test tubes.
2. Into each, by a graduated pipette, place 1 ml of normal saline.
3. Add 1 ml of serum in the first tube. Mix it well, by sucking and releasing, for 3 or 4 times.
4. Take 1 ml. from the first tube and place in the second tube. Mix as before. Then from this take 1 ml and place in the 3rd tube. this way make serial dilutions. So the dilutions will be 1 in 2 : 1 in 8 : 1 in 16 : 1 in 32 : 1 in 64 : 1 in 128 : 1 in 256.
5. To each tube add from a graduated pipette 0.25 ml. of 10% sodium carbonate and mix.
6. Then to each tube add by a graduated pipette 0.15 ml. of 0.5% mercuric bichloride and mix.
7. Read in half hour and after 24 hours at room temperature.
8. A positive reaction consists of dense flooculation, precipitated in at least the first 3 tubes.

This test is positive in animals showing cirrhosis and liver damage.

## ESTIMATION OF CLOTTING TIME

Prothrombin is required for clotting of blood and this prothrombin is manufactured in the liver. In diseased liver, this function is interfered with and so clotting occurs late. So the time taken for clotting of the blood can be used as a test to assess the health of the liver.

1. Insert a sterile needle in the vein, taking due precautions.
2. Start a stop watch when the blood begins to flow out of the needle.
3. Collect the blood, about 2 ml on a watch glass.
4. Plunge 6 to 8 capillary tubes 8 cms x 1 mm into the blood, when the blood rises into them by capillary attraction.
5. Keep them on a table and break small pieces of the capillary tube at half minute intervals. The end point is reached when a shred or string of clot is obtained.

The normal values for clottiog time by this method are;

Cattle — 6½ to 8 minutes
Horses — 1 1½ minutes
Sheep and goats — 2 minutes
Dogs — 2 minutes
Pigs — 3 minutes
Cats — 1 to 3 minutes

In cirrhosis in cattle, the clotting time was prolonged and was 5 to 17 minutes.

## VAN DEN BERGH TEST

For the theory behind this test see page 179 of the text book.

**Method**

Make the following solutions:

Solution A

| | |
|---|---|
| Sulphanilic acid | 1 gm |
| Conc. HCl | 15 ml |
| Distilled water to make | 1000 ml |

Solution B

| | |
|---|---|
| Sodium nitrite | 0.5 gm |
| Distilled water | 100 ml |

At the time of making the test, mix 10 ml of solution A with 0.3 ml of solution B. This is the Ehrlich's diazo reagent.

1. Place 1 ml of nonhemolysed serum from a fasting animal in a test tube.
2. Run down the sides of the test tube 0.5 ml of the diazo reagent so that it overlays on the serum.
3. In 30 seconds at the junction of the two solutions reddish-purple ring appears. This is the **direct test.**
4. If the color develops after 30 seconds it is **indirect test.**
5. Mix the contents and add 3 ml of 95% alcohol and mix. If there is pink color now and no color before adding alcohol it is **indirect test.** If there was pink color in the direct test and if it becomes intensified, then it is known as **biphasic.** Absence of any pink color is **negative test.**

The following table depicts the differences various types of jaundice may show when tested by laboratory methods.

| | Hemolytic (per-hepatic) | Toxic (Intra-hepatic) | Obstructive (Post-hepatic) |
|---|---|---|---|
| Serum-van den Bergh test | Indirect | Biphasic | Direct |
| Urine bilirubin | Not present | Present | Present |
| Urine urobilinogen | Slightly present | present | Not present |
| Feces | Intense yellow<br>No smell | Nornal | Clay colored<br>Greasy<br>Foul smell |
| Liver function test | Negative | Positive | Negative (in early stages) |
| Blood prothrombia time | Normal | Prolonged | Prolonged |
| Total serum cholesterol | Normal | Decreased | Increased |

## BROMSULPHALEIN (BSP) DYE EXCRETION TEST

One of the functions of the liver is the removel and excretion of foreign dyes from the blood. The dye is removed by the R.E. cells and excreted by the hepatic cells. In liver damage, this function is impaired and so measurement of the retention of a dye introduced into the blood gives an idea of the hepatic damage.

This test is most useful in dogs though it can be done in cattle also. Again this test can be conducted provided one has the necessary comparator and standards.

**Method**

1. Weigh the dog in pounds. Divide this by 55 to obtain the quantity of the 5% solution of dye in ml to be injected. The dosage obtained in 2 mg/kg body weight.
2. Inject the indicated quantity of the dye intravenously TAKE CARE TO SEE THAT NONE OF THE DYE SPILLS INTO THE SURROUNDING TISSUES
3. 30 minutes after the injection, remove 5 ml of blood from the vein other than the one used for injection of the dye, using heparin as anticoagulant.
4. Centrifuge the blood and remove 0.5 ml of plasma (which must be unhemolysed) into each of two small tubes supplied with the kit.
5. To one tube add 1 or 2 drops of 10% sodium hydroxide. To the other add a few drops of 5% HCI.
6. In the comparator, place the tube with the acid added (the blank) in front of the standard and the other tube with the alkali added is compared with the standard. On the standard tubes is etched the percentage, which denotes percentage of the dye retained.

Normally after 30 minutes only about 5% of the dye is retained in dogs with healthy livers But increased values denote hepatic injury.

BSP is retained in the following conditions:

1. Hepatic necrosis.

2. Hepatic fibrosis

3. Acute hepatopathies—focal necrosis, jipoidal retention, poisoning with $CCI_4^-$, infectious hepatitis, diabetes mellitus in which there is fatty infiltration, leptospirosis etc

4. In conditions where there is sluggish blood flow, as in cardiac decompensation, shock and servere dehydration (cardiovascular collapse). In these conditions the hepatic parenchyama need not be damaged.

**Serum Alkaline Phosphatase estimation** (see page 30). **SGPT activity** (see page 32)

## KIDNEY FUNCTION TESTS

To assess the function of kidneys, various tests have been devised, many of which are laborius and require specialised equipment The simplest and at the same time a reliable test is the measurment of the blood urea nitrogen. In kidney disease, the effete products of protein metabolism are not excreted and so are retained in the blood. Hence if the blood urea nitrogen content is measured, it gives us a fairly accurate idea of the health of the kidney. The normal blood urea nitrogen is 10 to 20 mg./100 ml of blood and values higher than this indicate renal damage.

**Method** The mercury combining power

**Principle** Mercury combines with urea when mercury salt is added to a substance containing it. When all the urea has combined with mercury, excess of mercuric chloride (which is used in this test) is free to combine with sodium carbonate, when the latter is added, to from red mercury carbonate, which is taken adantage of as the end point in the test.

**Procedure**

1. Collect blood with an anticoagulant other than ammonium oxalate or double oxalate.

2. Add drop by drop 5 ml of this blood to 5 ml of 10% trichloroacetic acid in a centrifuge tube (This is prepared by adding 10 gms of trichloroacetic acid to 50 ml distilled water. After it dissolves, make upto 100 ml with distilled water.

3. Mix well. Centrifuge the mixture or if a centrifuge is not available, filter.

4. To 5 ml of the supernatant fluid (or the filtrate) add froma burette 1.5 ml of 5 % mercuric chloride solution (Prepared by adding 5 gm of mercuric chloride to 50 ml distilled water and after it dissolves making up to 100 ml by distilled water).

5. Continue adding the mercuric chloride in small quantities till the end point is reached this is assessed as follows. A drop of the mixture in the test tube is placed on a glass place and to it is added a drop of staurated solution of sodium carbonate. If a reddish brown precipitate forms within 3 seconds, then end point is reached. So it is necessary to spot check often this end point while titrating with the mercuric chloride solution. A yellow precipitate that may form before the reddish-brown precipitate forms is not the end point.

6. Calculate the blood urea nitrogen as follow.

$$\frac{40 \times \text{no. of ml of } HgCl_2 \text{ used minus } 60}{21.4} = \text{the blood urea nitrogen in mg per 100 ml of blood.}$$

# 6

# CEREBROSPINAL FLUID

| | |
|---|---|
| Method of collection | None-Apelt test |
| Physical examination | Glucose |
| Color | Cytological examination |
| Turbidity | Diluting fluid |
| Coagulation | Counting of cells |
| Chemical examination | Pleocytosis |
| Proteins | Nature of CSF in diseases |
| Pandy's test | |

Examination of cerebrospinal flluid is undertaken:

1. Whenever diseases of CNS, basing on clinical evidence are suspected;
2. As a useful prognostic aid in the diseases of CNS;
3. To assess the response to treatment;
4. As a therapeutic measure to remove certain amount of CSF
5. To introduce substances for X-ray examination and
6. For introduction of anesthetics or drugs or serum.

**Methods of Collection of CSF**

Two sites are available for collection of the CSF. One is the puncture of cisterina magna (suboccipital). This site is most useful for horses, cattle, sheep, dogs and cats. The second is the sublumbar puncture. This is an alternative site useful in cattle, sheep and goats.

**Porcedure**

**Suboccipital pucture:** Cast the animal on the right side (if you are a right handed person). Flex the head to the left so that its longitudinal axis forms a right angle with the longitudinal axis of the neck. Shave, clean and sterilise the area. Take a 3 to 4 inches long, 16 gauge, sterile, needle with a stylet and insert it at the cervical midline at the level of cranial edge of the wings of the atlas. The direction of the needle should be parallel to the long axis of the head. When the needle enters the subarachnoid space, you suddenly feel no resistance. It is at this juncture that the head of the animal should be held firmly as it may be jerked suddenly as the puncture is made. On removal of the stylet, CSF will flow out. If ready flow does not occur, withdraw 5 ml of the fluid by means of a steriule syringe, ( Animal may have to be anesthetised).

**Sublumbar pucture:** In this procedure, CSF can be collected while the animal is standing. The site is a depression found in the midline between the dorsal spinous processes of the last lumbar vertebra and the cranial edge of the median sarcral crest. The area is shaved, cleaned and sterilised A long sterile 5 inch, 14 to 16 gauge, needle is inserted in the midline. on entering into the subarachnoid space, CSF will flow out when the stylet is withdrawn. If the fluid does not flow out, 5 ml. of it may be drawn with a sterile syringe.

Cell counts, tests for glucose and examination for bacteria, must be performed immediatly after collection of the CSF. Other tests can be delayed provided it is refrigerated.

## PHYSICAL EXAMINATION

**Color:** Normally, CSF is colorless and clear like distilled water, if it is bright red, then it may be due to puncture of blood vessels while inserting the needle. To make sure, centrifuge it. The supernatant, if colorless, confirms this assumption.

**Dull red or brown** coloration may be due to:

(a) Intracranial hemorrhage
(b) Chronic pachymeningitis or
(c) Fracture of cramium.

**Yellow color (xanthochromic)** may be due to presence of bilirubin, consequent on disintegration of red blood cells within the subarachnoid space. This may mostly be due to:

(a) hemorrhage in the subarachnoid space:
(b) alteration of the permeability of the lining membrane allowing, thereby, the entry of pigments of blood plasma into the CSF;
(c) Spinal block:
(d) neoplasms;
(e) abscesses;
(f) severe jaundice:
(g) carotememia;
(h) toxplasmosis;

**Gray or green** color may be due to suppurative conditions.

**Turbidity:** Normally CSF is perfectly transparent without any turbidity. Turbidity is usually due to presence of cells and this is evident, grossly, only if there are 500 or more of cells per cu. mm.

In acute bacterical meningitis, the CSF is cloudy or even purulent.

As already mentioned, hemorrhage due to puncture of blood vessels during collection gives rise to a red turbid fluid.

**Coagulation:** Normal CSF does not coagulate.

If the fluid contains abnormal amounts of proteins, especially fibrinogen, it may coagulate.

Presence of large quantities of blood, either due to internal hemorrhage or contamination during collection of the CSF may cause coagulation.

## CHEMICAL EXAMINATION

**Protein:** The normal protein content of the CSF is very small, being 12 to 40 mg./100 ml. This is almost wholly made up of albumin. Globulins having larger molecules do not pass into the CSF.

Globulins are the proteins of pathological interest. If CSF contains blood, then tests for globulin cannot be applied since serum globulins contaminate such a fluid.

### Quantitative Tests for Globulin

**Pandy's test:** To 1 ml. of saturated aqueous solution of phenol, add a few drops of CSF and shake the mixture. If turbidity develops, globulins are present (Normal CSF may produce only a very faint turbidity).

**Nonne-Apelt test:** Take 1 ml of filtered saturated aqueous solution. On this layer, carefully, 1 ml of the CSF. Keep for a few minutes. If at the juction of the two liquids a whitish ring forms, globulins are present. Normal CSF does not produce a ring.

**Interpretation:** Increased globulines in CSF indicate the

A. Presence of inflammation, as in the following conditions:
   (i) Encephalitis
   (ii) Abscesses in the brain and spinal cord
   (iii) Meningitis.
   (iv) Toxoplasmosis

B. Increase in the capillary permeability as in the following conditions:
   (i) Pneumonia
   (ii) Convulsions
   (iii) Uremia

C. Hemorrhages

D. Tissue destruction as in

**CATTLE**

(i) Congestion
(ii) Edema
(iii) Hemorrhage.

E. Neoplasms (occasionally)

A decreased globulin content of CSF is a favourable sign and indicates that the inflammatory conditionsis subsiding.

**Glucose:** Normally CSF contains 40 to 80 mg/100 ml.

The glucose level of CSF is dependent upon:

(i) the blood glucose level
(ii) the selective permeability of the blood-CSF barrier
(iii) the presence or absence of glycolytic bacteria.

**Method:** Method of estimation is similar to the estimation of blood glucose and must be done in fasting animals.

Increased gllucose levels—(hyperglycorrhachia) ocurs in:

(i) Hyperglycemia(diabetes mellitus especially).
(ii) Encephalitis
(iii) Compression of spinal cord
(iv) Brain tumours
(v) Abscesses of brain

(ii)–(v): Slight increase

Decreased glucose level (hypoglycorrhachia) occurs in:

(i) Hypoglycemia
(ii) Acute pyogenic infections.

## CYTOLOGIC EXAMINATION

The total cell counts in the CSF must be estimated within 20 minutes of collection, since cells degenerate rapidly. Normal fluid contains small number of lymphocytes and rarely mononulears, histiocytes and endothelial cells.

**Diluting fluid**

| | |
|---|---|
| Glacial acetic acid | 10 ml |
| Distilled water | 90 ml |
| Crystal violet | 0.1 gm |

**Method**

1. Use white cell pipette
2. Draw the diluting fluid upto1 mark.
3. Draw the CSF upto 11 mark.
4. Mix thoroughly.
5. Discard the first 2 or 3 drops.

6. Fill the chambers on both sides of the standard hemocytometer.
7. Wait for 2 minutes for the cells to settle,
8. Count the cells on all the squares on the 2 sides (that is on 18 sq. mm of the area)
9. Multiply the total number obtained by 0.6 to give the number of cells in a cu. mm. of the CSF.

**Normal Counts**

Cattle, sheep and pig 0 to 15 cells/cu. mm.
Dog---upto 25 cells/cu. mm.
Horse---upto 23 cells/cu. mm.

**Increased number of white cells—Pleocytosis occurs in:**

1. Inflammatory lesions or irritation of brain, spinal cord or meninges. In acute pyogenic menigitis, cell counts are very high above 500 cells/cu. mm.
2. abscess of brain or spinal cord.
3. encephalitis.
4. neoplasms.
5. Chronic inflammatory, toxic or degenerative conditions.

**Differential count:** If the total count is less than 500 cells/cu. mm. centrifuge the fluid and make a smear of the sediment. Dry in air, stain with Leishman's stain and examine. Normally only lymphocytes or occasionally mononuclears or histiocytes may be found.

**Neutrophils:** denote presence of pyogenic or bacterial infection and are found in

(i) abscesses,
(ii) bacterial encephalitis,
(iii) bacterial meningitis
(iv) hemorrhage

**Lymphocytes increase in**

(i) viral infections
(ii) fungal infections
(iii) chronic infections
(iv) uremia
(v) toxic conditions
(vi) postvaccinial encephalitis

**Bacteriological examination** of the CSF is indicated when he cell count and the protein content are high.

**Nature of CSF in different diseases**

The following data are obtained from "Comparative Neuropathology" by Innes and Saunders. (Contributed by R. Fankhauser).

**Tuberculosis:** The CSF has the following characters in tuberculosis opaque and sometimes xonthochromic with a fibrinous netwerk forming a 'spider-web' clot after standing for some time; hundred to several thousands of cells per cu. mm, neutrophils and lymphocytes, monocytes and endothelial cells are seen; sugar content varies between 27 to 59 mg/100 ml protein content is high; Bacteriologically *M. tuberculosis* can be cultured.

**Purulent meningoencephalitis** The CSF is opaque, whitish or yellowish with fibrin filaments or flakes, often forming a clot; normal to augmented pressure; cell content is 75 to several thousands per cu. mm of which neutrophils predominate; sugar 50 to 78 mg% and protein content is hgih. Bacteria can be cultured.

**Non-purulent meningoencephalitis.** The CSF is clear and colorless or occasionally slightly opaque. Pressure is normal; cells are 10 to 140 per cu. mm and these are mostly lymphocytes and monocytes. Protein is high.

## HORSES

**American equine ancephalitis:** increaqsed CSF; turbidiy is present and cells vary from 600 to 25000 per cu. mm.
**Borna disease:** Protein is high and cells are 10 to 33 cu. mm.
**Purulent leptomeningitis:** Yellowish and turbid, cells higher than 350 cu. mm, all being neutrophils and high protein content.

## SHEEP

**Chronic encephalitis:** Pleocytoisis 50 to 2000 per cu. mm, all of these being lymphocytes; high protein.

## PIGS

**Pollymyelitis:** Cells range from 5 to 1000 per cu. mm all being monocytes; sugar is 92 to 113 mg % and high protein.

## DOG

In toxic encephalopathies, in acute meningoencephalities and servre vascular lesions and hemorrhages there is xthochromic CSF with erythrocytes and high protein content.

In general,

1. Alteration of CSF is present in inflammation of central nervous system and meaninges. So there is increased number of cellsand higher protein.
2. There is some parallelism between the dgree of alteration in the CSF and the severity of histopatholigical picture.
3. Marked increase in cells and protein indicates severe meningitic lesion.

# 7

# CLINICAL EXAMINATION OF SEMEN

(By Dr. P. Narasimba Rao)

Physical characters
- Macroscopic characters
- Color and consistency
- Volume
- Density

Physiological, biochemical and viability studies
- Mass activity
- Individual morility of Spermatozoa
- cold shock
- Stroage motility
- pH of semen
- Fructolysis index
- Methylene blue reduction
- Resazurin reduction

Microscopic evalution of number of spermatozoa andd sperm morphology
- Serum concentration
- Counting live and dead sperms
- Staining by Eosin & Nigrosin
  - William's stain
- Head abnormalities
- Tail abnormalities
- Mid piece adnormalities
- Protoplasmic droplets presents

Examination of semen for genital infections.
- *Campylobacter fetus*
- *Brucellosis*
- Virus infections
- *Trichomonas fetus.*

Clinical examination of seman is very essential in the control of fertility in artificial insemination work and in the control of spread of coital diseases.

The assessment of morphological and physiological aspects of semen must be undertaken immdiately after collection. The viability of semen after processing and storage is also to be evaluated for successful implementation of A I. Even though the prefered method of collection for the evaluation is by artificial vagina, semen obtained by electro-ejaculation can also be used. Semen obtained by electro-ejaculation is good for the evaluation of motility and sperm morphology.

Semen is subjected to the evaluation of the following aspects: (A) physical. (B) Physiological, (C) Biochemical, (D) Storability and (E) Presence of infectious agents.

**Physical characters of the Semen:** The physical characters of the semen are presented in the following table.

**Comparative size of spermatozoa of farm animal and rooster.**

| | Head | | Middle piece | | Tail piece | |
|---|---|---|---|---|---|---|
| | Length (μ) | Width (μ) | Length (μ) | Width (μ) | Length (μ) | Width (μ) |
| Bull | 9.15 | 4.25 | 14.84 | 0.67 | 45-50 | 0.51 |
| Ram | 8.2 | 4.25 | 14 | 0.80 | 40-45 | 0.50 |
| Boar | 8.5 | 4.25 | 10 | - | 30 | - |
| Stallion | 5 | 2.4 | 8 | 0.50 | 30 | 0.49 |
| Rooster | 100 (Total length) | | | | | |

**Table of semen characteristics of domestic animals.**

| Species | Average Volume of ejaculate c. cs. | pH | Average concentration (millions per c. c.) | Storage survival time with retention of fertilising capacity (days) | Number of females that can be bred per ejaculate |
|---|---|---|---|---|---|
| Bull | 4 | 6.4 to 6.8 | 600-1200 | 4-6 | 100-600 |
| Murrah Buffaloe Bull | 2 | 6.2 to 6.6 | 800-1200 | 2-3 | 100-400 |
| Ram | 1 | 5.9 to 6.8 | 800-4000 | 1-2 | 40-12 |
| Stallion | 50-100 | 7.3 to 7.9 | 50-200 | -- | 8-12 |
| Boar | 200-250 | 7.0 to 7.9 | 25-1000 | Few hours to 24 hours(Fertility drops after 30 hours). | 4-8 |
| Cock | 0.1 to 1.5 | 7.02 to 7.18 | 3800 | Always use fresh | 6-10 |

## I Macroscopic Character

1. **Color and Consistency** This is usually estimated on visual examination immediately after collection and clasiffied as mentioned below:

(a) Creamy: thick or thin
(b) Milky; thick or thin
(c) Watery
(d) Cloudy

Creamy samples are commonly noticed in bulls, bucks and rams.
Milky samples are commonly seen in buffaloes.

2. The VOLUME of semen can easily be recorded in milliliters directly from the semen collecting tube.

3. DENSITY is only a rough estimation of numbers of spermatozoa carried out visually and can be compared by microscopical counting.

D = 400 millions/ml of semen.
DD = 800 million/ml of semen
DDD = 1200 million/ml of semen

Intermediate stages are sometimes described as

(D) = 200,000
D(D) = 600,000
DD(D) = 1000,000
DDD(D) = more than 1200,000

## II Physiological, Biochemical and Viability Studies

1. **Mass activity:** Immediately after collection take a small drop of semen and examine it under low power without a cover slip on a biotherm. The mass activity is commonly described in terms 0 to +++

0=no waves;+=very low waves; ++ moderate waves without plenty of whirlpools; +++=rapid and swift waves with whirpools.

2. **Individual motility of spermatozoa:** The estimation of individual motility of spermatozoa for each collection made should be arrived at as a matter of of routine. This is the best practical indicator to tell us the quality of seemen used for insemination work.

Immediately after collection, one drop of the semen is diluted (1in 100 roughly) with Ringer'ssolution at 37°C. A droplet of semen is transfered to a clean slide kept warm on a blotherm at 37°C and covered with cover slip to make a thin unform semen film. The preparation is then examined underr high power. The number of sperms exhibiting progressive motility are roughly estimated against the total number of sperms in a given field. A few fields are studied and assessed. The following is the notation used: 1 = 20%; 2 = 40%; 3 = 60%: 4 = 80% (4) = more than 80% progressively motile sperms.

3. **Cold shock** is an expression of tolerance of sperms to sudden variations in ambient temperature.

As soon as semen is collected. determine the percentage of live sperms by differential stain. Then immediately immerse 1 ml of semen sample (preferably diluted) in a water bath at 0°C for 10 minutes. Then determine the percentage of live sperm after the cold shock. The percentage of sperms remaining alive after shock may be expected to maintain motility for 2 to 5 days under proper storage conditions.

4. **Storoge motility** is an expression of motility at intervals of 24 hours stored at the appropriate temperature for that particular dilutor. The number of hours a semen sample maintains its viability without the motility falling below 3 is a good index of semen viability.

5. **pH of semen** is measured by using a pH indicator paper or by a pH meter. Capillary pH indicators are also used.

6. **Fructolysis index** is defined as the amount of fructose utilised by $1x10^9$ cells per hour. For normal bovine semen the rate is 1.2 to 2.9 mg of fructose per hour. The technique demands a biological oxygen demand incubator. The index depends on a number of factors such as time that elapses following ejaculation, the availability of oxygen and glucose.

7. **Methylene Blue reduction** Oxygen is used at a more rapid rate in semen containing high concentration of active sperms than in poor quality semen. This causes an excess of production of hydrogen ion concentration which combines with the methylene blue to form leuco-methylene blue. The time reuired for this reduction of methylene blue color serves as an indicator as to the number of active sperms present in semen. The rate of reduction in semen depends largely upon the concentration and the activity of spermstozoa. However , it may also be influenced to a lesser extent by such factors as bacteris, enzymes and light but the time taken for reduction by these agents is much larger and hence they are excluded from this method.

**Method:** (i) Prepare methylene blue solution by dissolving 50 mg, of methylene blue in 100 ml of citrate buffer.

(ii) Dilute 0.2 ml of semen with 0.8 ml of egg yolk citrate dilutor in a 10 ml vial and mix.

(iii) Add 0.1 ml. of methylene blue solution and mix.

(iv) Seal the mixture by layering with liquid paraffin, 1.0 centimeter thick.

(v) Observe the time required for the sample to lose its blue color.

The semen which bleaches the blue color within 9 minutes is considered to be of good quality.

8. **Resazurin reduction:** Resazurin, a hydrogen acceptor which changes color upon reduction replaces methylene blue in the above test.

## Microscopoic Evaluation of Number of Spermatozoa and Sperm Morphology

1. **Sperm concentration** The number of sperms per ml of semen can be counted by using a hemocytometer. Semen is diluted to 1:1000 with 3% salien (with eosin) 0.1 ml of

semen is first diluted with 9.9 ml of saline. Further dilution is made by ading 1.0 ml of this diluted semen to 9.0 ml of saline. the hemocytometer is charged with this diluted semen and the number of sperms present in 16 large squares, are counted. If A is the number of sperms present in 1 ml of semen is A x 10 millions.

A quick method of estimation of sperm concentration is by comparison with standard opacity tubes containing barium sulphate suspension of different concentrations.

The most commonly used method of assesment of sperm concentration is estimation of absorption of light by using a photo electric colorimeter. The semen is diluted 1 in 100 with isotonic citrate solution and the absorbency of light is read taking blank citrate solution as 100% transmission. Initially the absorbency and the actual concentration of sperms obtained by using a hemocytometer are plotted as a graph. Once the absorbency is obtained the number of sperms/ml are directly read from the graph.

An electronic Coulter counter can also be used for counting the number of sperms/ml of semen. But this apparatus is very costly.

**2. Counting Live and Dead Sperms** (Vital staining)

**Eosin and Nigrosin :** One drop of semen is mixed with 2 drops of 5% aqueous eosin and one drop of 10% aqueous nigrosin. The entire quantity is thoroughly mixed and a thin film is made from the mixture. Live sperms appear unstained and dead sperms are stained pink against a brownish-purple background. Nigrosin is not an easy stain to make and so India Ink may be substituted for nagative staining.

**Staining for Study of Morphology**

William's method of staining is popularly used for the study of sperm morphology.

**Composition of William's Stain**

Stock solutions:
A. 10% Basic Fuchsin in 96% alcohol.
B. Saturated solution of Eosin (blue) in 96% alcohols. Carbol fuchsin stain. Add 10 ml of stock solution A to 90 ml of 5% aqueous phenol solution.

William's stain is a mixture oft equal parts of saturated eosin blue solution and carbol fuchsin and repened for 14 days.

1. Dilute the semen to facilitate drawing a smear of individual sperms.
2. Air dry and fix in flame.
3. Further fix it, 3 to 4 minutes, in absolute alcohol.
4. Treat for 1 to 2 minutes with 1/2% chloramine to remove mucus
5. Wash the smear with water and then with absolute alcohol
6. Stain with William; stain for 7 to 10 minutes.
7. Wash the slide with water and dry. The sperms are observed under oil immersion and abnormalities are classified as follows.

A. **Head abnormalities:** 1. Narrow, 2. Narrow at abase. 3. Pear shaped 4. Abaxial, 5. . Round, 6. Macro; 7. Micro. 8. Broad. 9. Weakly stained, 10. Under developed, 11. Detached head.

B. **Tail abnormalities:** (a) bent tails (b) coiled tails.

C. **Mid piece abnormalities.**

D. **Protoplasmic droplets present**

The total number of abnormal sperms should not exceed 16 %.

## Examination of Semen for Genital Infections

Presence of microorganism like *Mycobacterium tuberculosis, Brucella abortus, Corynebacterium pyogenes Streptococci and Staphylococci* can be demonstrated by culture methods. *Compylobactor fetus* can also be demonstrated by culure methods, even though the procedure is cumbersome.

*Campylobacter fetus.* Contaminants can be overcome to a large extent, by using microfilter before inoculation of media. Primary cultures must be incubated under 10% $CO_2$ or in a gas mixture containing not more than 5% oxygen. Any conventional media can be used even though numerous special madia are available.

The commonly employed method for diagnosis of vibriosis is to test.mate virgin heifer and test for presence of antibodies for *C.* fetus in the vagina mucus.

**Brucellosis:** Agglutination test on seminal plasma can be conducted for detection of antibodies against *Brucella abortus* by the conventional tube method. However, serum agglutination method is more sensitive.

**Viral infectious agents:** The presence of viral infectious agents in semen can be established by tissue culture techniques.

**Trichomonas fetus:** Presence of *Trichomonas fetus* in the semen can be-detected by microscopic observation of moving protozoa. Also cultures from preputial washings can be obtained.

# 8

# PREGNANCY DIAGNOSIS

By Dr. P. Narasimha Rao


| | |
|---|---|
| Manual examination | Immunological test |
| Rectal examination | Pregnancy diagnosis using ultrasonic equipment |
| Abdominal ballottement | |
| Biological test | Radioimmunological assay |
| Chemical test | Radiographic methods |


## MANUAL EXAMINATION

**A: Rectal examination:** Pregnancy can be diagnosed accurately by rectal examination in mare and cow from about 40th day onwards. Pregnancy is diagnosed by palpating the following 1. presence of corpus luteum, 2. dissimilarity of uterine horns. 3. fluidy fluctuation of the uterine horns, 4. presence of amniotic vesicles, 5. presence of double membranes (in the cow) : 6. presence of placentomes (in cow) ; 7. enlargement of uterine horns and 8. displacement of the uterus.

**B: Abdominal ballottement:** Pregnancy in sheep and bitch can be diagnosed by this method. However results of diagnosis are not accurate.

## BIOLOGICAL TEST

This test is useful only in the mare, in which after the 49th day of pregnancy, in the serum is present sufficient pregnant mare serum gonadotrophin (PMSG) which can be detected by this test. This test is useful between the 49th and 140th days of pregnancy.

**The Test**

1. Collect blood and separate the serum.
2. Inject 10 ml of the serum into the marginal ear vein of a mature, non-pregnant female rabbit, which has been isolated from a male rabbit for a t least 30 days.
3. After 48 hours, sacrifice the rabbit and examine the ovaries. In a positive case, dark red follicles, *corpora hemorrhagica*, are seen.

Since rabbits are costly, this test is performed in rats. The following procedure is followed.

1. Take two or three immature female rats aged 22 days.
2. Inject 5 ml of serum from a suspected animal, into the rats subcutaneously.
3. Kill the rats 96 hours later.
4. Presence of hemorrhagic spots (corpora hemorrhagica) is positive of pregnancy.

## CHEMICAL TEST

This test is dependent upon the presence of estrogens in the urine of pregnant mares and is useful only between the 120th day of pregnancy and to the end of term. In this test, presence of hormones or their congugates in the serum of urine can be detected by the development of color by the interaction of the hormones or their derivatives with concentrated sulphuric acid. This is know as Cuboni test and is useful only in the mare and pig.

**The Cubnoi Test**

1. Filter the Urine and take 5 ml in a test tube containing 1 ml of concentrated HCl.
2. Place the tube in boiling water bath for 10 minutes.
3. Cool the tube under running water.
4. Add 6 ml of benzene and shake vigourously to obtain thorough mixing.
5. Allow the benzene to separate.
6. Take 3 ml of the extract in a dry test tube and dry it at 60°C to 80°C.
7. To the residue add 4 ml of concentrated sulphuric acid.
8. Warm this on a water bath at 70°C to 80°C for 10 minutes and shake the tube several times while heating.
9. Stopper the tube and then shake vigourously and then examine it in a dark room under reflected light for the presence of green fluorescence which is indicative of a positiv test.

**Sows**

Usually it is not necessay to resort to chemical tests in the sow. The Cuboni test can be performed between 21st and 32nd days after conception and again between 72 days to parturition.

## IMMUNOLOGICAL TEST

A test kit for pregnancy diagnosis in the mare is available from Denver Chemical Manufacturing Co. Stanford. Connceticut, USA. (Detailed description of the test is available in the article. Immunological pregnancy test for Mares by Solmon WJ and Hoff. G (1969). J.Am. Vet. Med. Assoc, 155. 42).

The test is called Hemagglutination Inhibition test. It is based on the presence of Pregnant Mare Serum Gonadotrophin (PMSG) in pregnant mare between the 40th and 120th days of pregnancy. Sheep red blood cells are coated with PMSG. Test kits generally have latex particlcs coated with PMSG. The antibodies added to the test serum are neutralized by the PMSG present in the positive test serum. So the PMSG coated RBCs settle slowly to the bottom of the test tube, as a ring. The coated RBCs with the negative test serum agglutinate with PMSG antibody and settle fast to the bottom of the test tube as a layer. The antibody is obtained from rabbits (serum) immunised with PMSG. Neutralizer against non specific serum agglutinins is also added while conducting the test.

## PREGNANCY DIAGNOSIS USING ULTRASONIC EQUIPMENT

In sheep and sows, pregnancy can be diagnosed by passing ultrasonic waves and receiving them back. The time lapse in between sending and receiving the pulse gives a clue for pregnancy. The ultrasonic Doppler instrument is inserted per rectum. A high accuracy in the diagnosis is reported.

## RADIOIMMUNOLOGICAL ASSAY FOR PMSG

This test is very complicated and though accurate, can be conducted only in a well equipped laboratory. This is mentioned here to kindle the enthusiasm and curiosity of interested students.

Radioactive iodinated PMSG (Hot) is agglutinated with antibody PMSG and gamma globulin from heterologous species. The Hot PMSG is replaced by using the principle of competitive binding with known quantities of uniodinated hormone (cold). The hot hormone is separated from the bound complex and radio activity is measured on a gamma counter. A standard graph is prepared. The PMSG from the test is used now to replace the hot PMSG and decrease in radioactivity is measured. This gives an accurate measure of serum level of PMSG.

## RADIOGRAPHIC METHODS

X-rays are used for diagnosis of pregnancy in dogs and cats. The calcified fetal skeleton is radio-opaque and ao can be easily seen . In the bitch this is useful to diagnose pregnancy after 45th day. Radiographic diagnosis of pregnancy is possible in ewes also, from the 70th day of pregnancy onwards.

## OVA OF HELMINTHS FOUND IN FECES OF ANIMALS AND FOWLS

1. Oxyuris equi (Horse) With a plug at one end.
2. Spirocerca lupi (Dog) Gelatine-capsule shaped with larva inside.
3. Taenia taeniaformis (Cat) Hexacanth-embryo (six hooked) inside.
4. Ancylostoma caninum (Dog) Oval egg with thin shell.
5. Dipylidium caninum (Dog) Packet of eggs and each egg has a hexacanth embryo.
6. Esophagostomum columbianum (Sheep).
7. Fasciola gigantica (Cattle, sheep, etc) Large operculated, yellow-tinted.
8. Schistosoma japonicum (Cattle) An oval egg with a terminal spine.
9. Echinococcus granulosus (Dog) Egg with hexacanth embryo.
10. Diphylloborthrium latum (Dog) An operculated egg.
11. Ascaris lumbricoides (Pig) Thick pitted shell.
12. Dicrocoelium dendriticum (Sheep) Operculated egg.
13. Dictyocaulus arnifieldi (Horse) Strongyl egg with larva inside.
14. Strongyloides papillosus (Sheep) Egg with a larva inside.
15. Capillaria sp. (Fowls) Egg with plugs at both ends.
16. Trichuris ep. (All animals) Egg with plugs at both ends.
17. Syngamus laryngeus (Cattle).
18. Dioctophyma renale (Dog) Egg with every thick pitted shell.
19. Macracanthorhynchus hirudinaceus (Pig).
20. Amphistome sp. (Cattle, Sheep, etc.) Operculated egg.
21. Schistosoma suis (Pig) Egg with a spine at one end.
22. Schistosoma nasale (Cattle) A boomerang shaped egg with spine at one end.
23. Ascaridia galli (Fowl).
24. Heterakis gallinae (Fowl).
25. Physocophalus sexalatus (Pig).
26. Mesocestoides Ilneatus (Dog).
27. Toxascaris leonina (Dog) Egg with finely pitted thick shell.
28. Trichostrongyle sp. (Cattle , sheep, etc.)
29. Gnathostoma spinigerum (Cat).
30. Toxocara canis (Dog) Egg with coarsely pitted thick shell.
31. Ascaris vitulorum (Cattle) Pitted thick shell.
32. Schistosoma spindale (Cattle) Elongated egg with spine at one end.

# OVA OF HELMINTHS FOUND IN FECES OF ANIMALS AND FOWLS

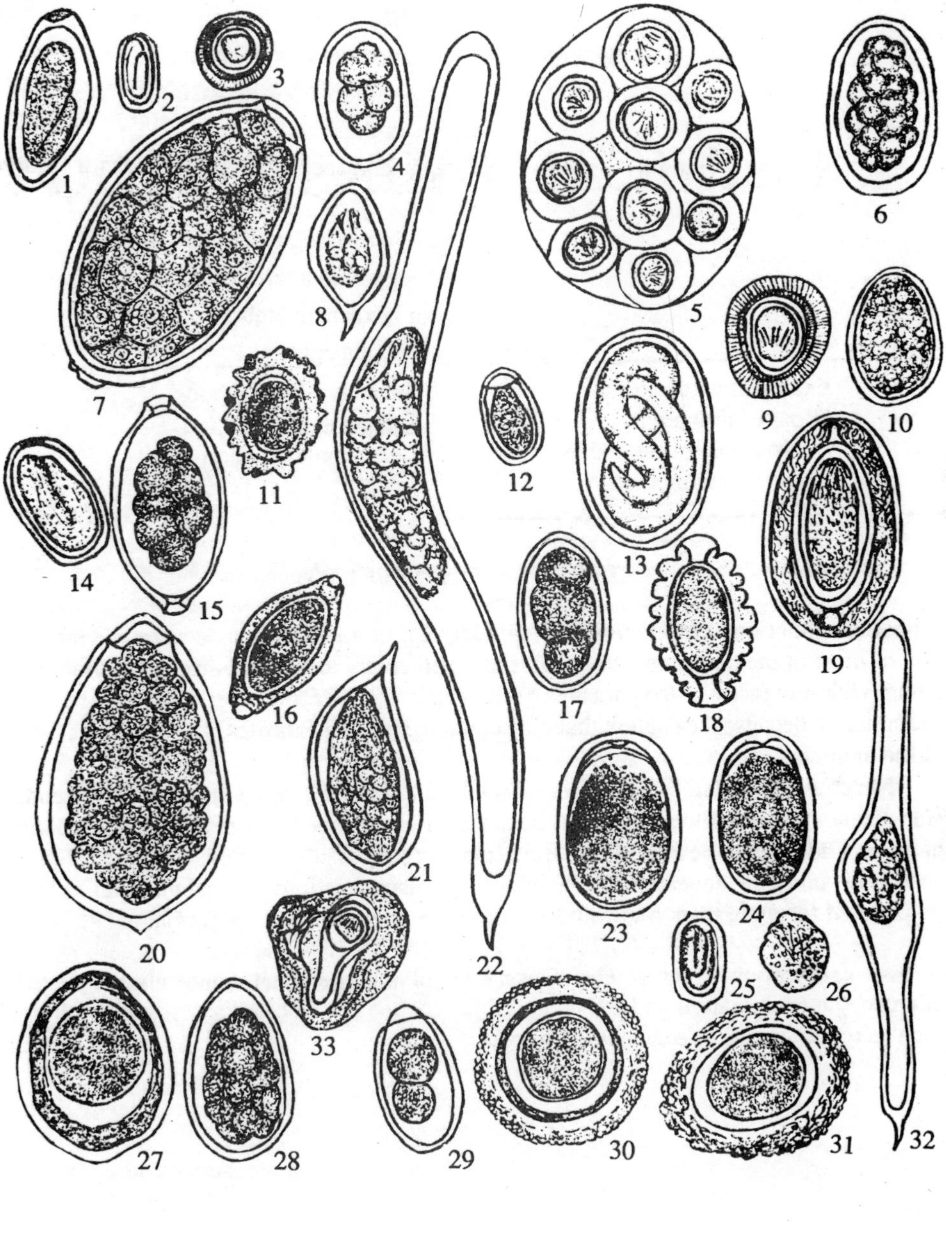

# 9

# CLINICAL PARASITOLOGY

| | |
|---|---|
| Examination of feces | Concentration methods |
| Gross examination | Sedimentation method |
| Microscopic examination | Floatation methods |
| Direct semar method | Examination of skin scrapings |

## EXAMINATION OF FECES

Proper examination of feces provides evidence for or, at times, an accurate identification of most of the parasites that inhabit the alimentary canal and discharge their eggs along with the feces. Also parasites of the respiratory tract can be diagnosed by fecal examination because the sputum that may contain eggs and parasites is swallowed in most of the animals.

Material submitted for the examination should be fresh and sufficient for dependable diagnosis. The collection should be made in clean containers without contamination of urine. Old samples become dehydarated making suspension difficult. Also eggs of parasites may undergo development, hatching or disintegration. If fecal material is to be transported for more than a few hours, it must be preserved in 10% formalin.

**Gross examination of feces:** Gross inspection of the specimen should always precede microscopic examination and the material be looked for macroscopic parasitic objects such as round worms or segments of worms.

**Appearance of various fecal materials:** Before examining the fecal samples for the eggs of the worms, one should know to which class of animal the sample belongs.

The dung of herbivora will be dark greenish-yellow with a peculiar smell and will contain lot of vegetable fibre. Beside fibres, protein grains, plant hairs, mould spores and animal debris that resemble parasitic forms (pseudoparasites) may be found.

The dung of carnivores will be soft and pasty, dark yellow in color, devoid of fibres, having unpleasant smell.

Parasitic eggs or cysts of one species may be found in the feces of another species of host, as a result of coprophagy (spurious parasites).

**Microscopic examination of feces:** This may include several techniques, such as direct examination and concentration methods.

**Direct semar examination:** A small quantity of the feces, about the size of a pin head is placed on a slide along with 2 or 3 drops of water, thoroughly emulsified with a needle, evenly spread over the slide and examined under the microscope. This method has many disadvantages and is not an accurate technique. It is useful when the infection is heavy.

**Concentration methods:** The rationale of the concentration is to separate parasitic objects from the bulk of the material in the specimen. For this purpose two methods, the sedimentation and floatation or a combniation of the two are employed.

**Sedimentation:** This method is reliable for all types of parasitic eggs.

**Procedure:** A small quantity of feces made up from different areas of the sample is thoroughly mixed with 10-15 ml of water, the emulsion is strained through a sieve to remove all coarser particles. This filtrate is poured into a centrifuge tube just upto one inch below the brim and centrifuged at 100 rpm for 1 to 2 minutes. All the eggs get packed at the bottom in the sediment and so a drop of the sediment is placed on a slide, covered with a cover glass and examined under the microscope.

**Floatation:** This methods is mostly useful in the examination of nematode infection only since eggs of trematodes, most of tthe cestodes and a few nematodes cannot be floated. Light infection can invariably be detected. The principle is to utilise a liquid suspending medium that is heavier than the parasitic objects, so that they rise to the surface and can be skimmed out of the surface film. For diagnositic usefulness the suspending medium must not only be heavier than the object to be floated but must not produce shrinkage to render the object undiagnosable. The common suspending media in use are:

1. Saturated solution of common salt (sp.gr.1.18-1.19).
2. Zinc sulphate 33% soln. (sp. gr. 1.180)
3. Saturated sugar solution (sp. gr. 1.25)
4. Saturated solution of sodium nitrate.

**Willi's technique:** Fill up to one third of a small floatation tube with thick emulsion of feces and then fill up the tube to its capacity with saturated solution of common salt to the brim till a convex surface is formed. Allow it to stand for half an hour by which time all the eggs would have floated up. Apply a clean coverslip or a slide to the surface of the fluid, removing just a drop of the fluid and this drop from the topmost layer is ecamined for eggs.

**Lane's method** (Centrifugal floatation)

A small quantity of feces is taken, emulsified in ten times of its own volume of water, strained through a sieve and centrifuged. The supernatant fluid is thrown off, the sediment is again suspended in water and centrifuged. Repeat centrifugation until the supernatant fluid is clear. Then the sediment is suspended in saturated sodium chloride solution and again centrifuged. Examination of a few drops of the fluid from the top-most layer will reveal the eggs.

**Zinc Sulphate Centrifugal Floatation Techinque**
**(For cysts and eggs exclusive of trematodes)**

**Procedure:** Prepare fecal suspension by comminuting about 10 parts of lukewarm water with one part of stool specimen. About 10 ml of the suspension is strained through a sieve and centrifuged for 45 to 60 seconds at 2300 rpm. Repeat centrifugation with water as in Lane's method until supernatant fluid is clear. Pour off the last s upernatant fluid. Add zinc sulphate solution and centrifuge at top speed for 45 to 50 seconds. Transfer several loopfuls of the material at the surface to a slide and examine for eggs etc.

Almost all cestodes, except the members of the pseudophillids do not discharge eggs but segments containing numerous eggs are shed and passed in the feces. Hence it is not always possible to find cestode eggs in fecal samples. A sizable quantity of feces is placed in a petri dish and the whole mass is searched for tapeworm segments with the aid of two mounted needles and hand lens if needed. The eggs of Fasciola sp. the large liver fluke look yellowish because of bile tint.

By the above methods the eggs of some worms can be recognised on account of their shape, size and other features and this fact is of great importance in making specific diagnosis.

The feces may also contain larvae, (e.g. some lung worms). Such larvae will be found by the concentration methods (floatation) used for detecting the presence of eggs.

Negative results need not necessarily mean that the animals are worm free, since the finding of eggs in feces is dependent on many factors.

## EXAMINATION OF SKIN SCRAPINGS

A sample of skin scraping is boiled with 5 ml of 10% NaOH or KOH solution in a test tube till a homogeneous suspension is got. Pour this into a centrifuge tube and centrifuge for 2 minutes at 1000 rpm. Examine a drop of the sediment under microscope for the mites or fungi (like ring worm).

**Normal Hematological Data of Ducks**

Erythrocytes (millions/c. mm) 2, 15 to 3.65
Hemoglobin (gms/100 ml) 11 to 16
PCV (%) 34 to 46
ESR (mm/hour) 1 to 2
Sp. Gr. 1037 to 1043
Fragility .22 to .46
White blood cells (thousands/c.mm) 32.56 to 60.0
Heterophils (%) 40 to 55
Lymphocytes(%) 23 to 48
Monocytes (%) 5 to 11
Eosinophils (%) 4 to 11
Basophils (%) 0 to 4.

Males have slightly higher values than females in respect of RBCs MB. PCV and total WBCs.

**Average Diameter in μ**

RBCs 12.10 x 6.45; Heterophilsi 9.84; Eosinophlis : 9.2; Lymphocytes, 8.9; Monocytes: 10.5; Basophils: 8.06.

# 10

# DIAGNOTIC TESTS

(Revised by Dr. S. Krishnaswamy)

---

- Staining procedures
  - Gram's stain
  - Ziehl-Neelsen's method
  - Leishman's method
- Tuberculosis in animals
- Tuberculosis in fowls.
- Johne' Disease
- Glanders (IDP test)
- Epizootic lymphangitis
- Pullolum disease
  - Rapid stained antigen
  - Tube test
- Direct Immuno peroxidase test for the diagnosis of Virus diseases
- Brucellosis
  - Modified Ziehl-Neelsen's stain
  - Stained antigen
  - Rose Bengal plate agglutination test
  - Tube test
  - Mercaptoethanol test
  - Milk ring test
- Rabies
  - Staining methods
- Mucus agglutination test for bovine vibriosis
- Agglutination test for avian mycoplasmosis
- Fluorescent antibody technique
- Surface fixation test for listeriosis

---

## Staining Procedures

### Gram's Method

1. Fix the slide by passing the side that does not contain the smear over a spirit lamp flame or bunsen burner a few times.

2. Cover the slide for half a minute with a 2% solution of crystal violet in methyl alcohol of highest purity (acetone free). If crystal violet is not available carbo-gentian violet can be used.

3. Pour off the stain and replace by Gram's iodine for half a minute.
4. Wash in abolute alcohol till purple color ceases to come off.
5. Wash in water.
6. Counter-stain with 1% Eosin solution for 1 minute,
7. Wash in water, dry and examine under oil immersion.

**Preparation of Carbo Gentian Violet Solution**

| | |
|---|---|
| Gentian violet | 1 gm |
| Carbolic acid crystals | 2 gm |
| Absolute alcohol | 30 ml |
| Distilled water to make | 100 ml |

**Gram's Iodine Solution**

| | |
|---|---|
| Iodine crystals | 1 gm |
| Potassium iodide | 2 gm |
| Distilled water | 300 ml |

**Results:** Gram positive bacteria stain violet. Gram negative red.

**Ziehi-Neelsen's Method or Staining by Acid-fast Method**

This method is useful to stain tubercular bacilli and Johne's bacilli.

1. Fix a dried smear by heat.
2. Cover the smear with carbol fuchsin.
3. With spirit lamp heat under the slide till steam comes off.
4. Allow the stain for 5 minutes.
5. Wash thoroughly with water.
6. Decolorise with 20% sulphuric acid
7. Wash in water and decolorise once again with acid.
8. Wash and repeat the acid treatment till only a faint pink color of the slide remains.
9. Wash throughly with water.
10. Decolorise with absolute alcohol for 2 minutes.
11. Wash in water.
12. Counter stain with 1% aqueous methylene blue for 1-2 minutes.
13. Wash in water, dry and examine in oil.

**Result :** Acid fast organisms take a red stain. Non-acid fast stain blue.

**Carbol Fuchsin Stain**

| | |
|---|---|
| Basic fuchsia stain powder | 1 gm |
| Carbolic acid crystals | 5 gm |
| Absolute alcohol | 10 ml |
| Distilled water to make | 100 ml |

### Preparation of Leishman's Stain

Take 0.15 gm of Leishman's stain powder in a glass mortar. To it add 10 ml of methyl alcohol (acetone free), triturate well with glass pestle and decant into a glass bottle. Still stain powder will remain. Add further quantity of methyl alcohol (10 ml) triturate and decant into the bottle. In this way triturate with 10 ml of methyl alcohol every time till all the powder dissolves. Finally to the bottle of stain add methyl alcohol to make a total quantity of 100 ml of the methyl alcohol. It is very important to triturate the powder of the stain with small alliquots of the methyl alcohol, otherwise all the powder will not get dissolved and unless all the powder is dissolved, you do not get a good stain and the smears will not stain well.

Method of staining blood smears by Leishman's stain has already been described (Page 16).

### Preparation of Stock Giemsa Stain

Dissolve 0.85 gms of the Giemsa stain powder in 100 ml of equal parts of glycerol and acetone free methyl alcohol by shaking and keeping for 4 or 5 days. A little deposit at the bottom does not interfere with staining capacities. Leishman-Giesma method of staining blood smears is described on page 16.

## TUBERCULIN TEST

The dobule intradermal (DID) test is used. For this tuberculin supplied by the Indian Veterinary Research Institute is utilised.

1. In the middle of the neck, on the left side, clip an area of skin 6 inches.
2. Clean it with alcohol.
3. Take tubercluin in tuberculin syringe which will deliver 0.1 ml at one plunge. If this is not availbale ordinary hypodermic syringe can be used. Take into this suring 0.1 ml of the tuberculin.
4. With the left fore finger and thumb grasp the skin tightly in the middle of the clipped area.
5. If you have calipers, you may measure the thickness of the skin here.
6. After holding the skin tightly inject into the deeper layers of the dermis (Note; it is not subcutaneous) the tuberculin.
7. Examine after 48 hours. If you find a diffuse, hot painful swelling then it is positive for tuberculosis. But if no diffuse swelling is present but only a small button is present then follow the next step.
8. Inject at the same place where you gave the first injection, 0.1 ml of tuberculin. Examine at the end of 72 hours.

Now in a positive case there will be a hot painful and edematous swelling. If this is not present, then it is negative. If a calipers is available, on masurement the skin will be double the normal thickness in a positive case. But the thickness of the skin alone is not the criterion. The character of the swelling is more important.

If stored in the refrigerator, the tuberculin will keep for 2 years. If it turns turbid, discard.

Recently a kaolin agglutination test using kaolin particles coated with phosphatide fraction from tubercle bacili is used to detect antibody in the sera of cattle. This is very specific.

**Tuberculin Testing in Birds**

In birds avian tuberculin supplied by I.V.R.I. is injected into the wattle. The other wattle serves as a control 0.2 ml. of the tuberculin is injected intradermally into the wattle. In a positive case, the wattle becomes swollan, hot and painful, in 48 hours.

If stored in the refrigerator, the avian tuberculin will keep for 2 years. If it turns turbid discard it.

## TEST FOR JOHNE'S DISEASE

**DID test with Johnin** This is conducted exactly as the DID test for tuberculosis. And the readings are also similar. A positive reaction consists of a diffuse, hot painful swelling after the second injection. The dose for each injection is 0.1 ml of Johnin supplied by I.V.R.I. An increase in the thickness of skin by 3 mm or more is positive.

If stored in the refrigerator, the Johnin will keep for 2 years. If it turns turbid discard it. Tuberculin from avain tubercle bacilli can also be used for DID testing as it cross reacts with *M. johnei*.

**Intravenous Johnmin Test**

This test is claimed to be more reliable than the serological and DID tests.

The test

1. Conduct the test when the environmental temperature is below n87°F.
2. Take the rectal temperature.
3. Then inject 2 ml of the johnin into the jugular vein.
4. Take rectal temperaturs at 4 1/2, 6 and 8 hours after the first injection.
5. The case is positive if the temperature is above 1.5° F over the preinjection temperature and the highest temperature and the highest temperature recorded is over 103° F.

A lymphocyte blast formation test using phyto hemagglutinins (PHA) on the purified lymphocytes from jone's Disease is reported to tbe diagnostic. This is based on the increased T-cell population in circulation. In the case of J.D. positive animals the lymphocyte actively multiples when stimulated with PAH. This is studied by increased uptake of labelled thymidine

## TEST FOR GLANDERS

**Intradermo-Palpebral Mallein Test (IDP)**

1. Take 0.2 ml of mallein (Suplied by IVRI) into a syringe.

2. If you are right handed person inject this i ntradermally into the lower eye lid, 1/2 inch below the eyelashes, about 1/2 inch from the inner canthus of the left eye. If you are a left handed person you can do this in the right eye.

3. In a glandered animal the reaction commences from about the 12th hour and it reaches its maximum by 24th hour and persists for 2 or 3 days. The eye becomes almost closed due to edema of the lower eye lid as well as other structures round about. The edematous part is hot and painful. Besides congerstion of the conjunctiva, there is a mucopurulent exudate from the inner canthus and the eyelids may become glued with an yellow exudate. In a negative animal there may be but a slight edema which clears off in a few hours. Temperature may be elevated, highest point being reached by the 12th hour.

If stored in a refrigerator, the mallein will keep for 2 years. If it turns turbid, discard it.

## EPIZOOTIC LYMPHANGITIS

A DID test can be done using Histoplasmin (cryptococcin). The test and the reactions are similar to those described for tuberculosis. A positive reactor show hot, painful and edematous swelling.

## PULLOTUM DISEASE

### Rapid Stained Antigen Whole Blood Test

1. Make a loop witha platinum or nichrome wire so that one loopful will be about 0.02 ml.
2. Puncture the wing vein and when blood oozes, dip the loop and remove one loopful and place on a glass plate.
3. Drop from a dropper one drop of the colored antigen. The drop should be 0.05 ml.
4. Mix with a broom stick and spread out for an inch in diameter.
5. Tilt the glass plate up and down to mix.
6. Reaction occurs within a few seconds to 2 minutes.
7. Delayed reactions are non-specific.
8. A positive test consists of the violet colored red cells becoming clumped together and floating in a clear fluid .
9. Only those rection in which clumping occurs within 1 minute of the mixing of the antigen and blood should be considered as positive. Those above 1 minute and within 2 minutes are to be taken as suspicious.
10. Immediately after depositing the blood, the loop should be agitated in distilled water and thus cleaned and then dried with a blotting paper.

This test is very rapid and useful to be conducted with large flocks. A table light may be used to light the glass plate to give uniform results. The glass plate should be clean and free from dust.

The antigen is a suspension of pure smooth culture of a standard strain of *salmonella pullorum* in a solution containing 1 % formalin, 1% $KH_2PO_4$ and 0.85% NaCl.

The antigenshould be kept at a temperature of 2°C to 5°C . It keeps for 3 months.

## TUBLE TEST FOR PULLORUM DISEASE WITH PLAIN ANTIGEN

The antigen as supplied by the IVRI is a suspension of pure smooth culture of Salmonella pullorum in phenol saline.

### Procedure

1. Keep 5 agglutination tubes in a rack.
2. Pipette 0.8 ml of carbol saline in the first tube (carbol saline is prepared by making a 0.5% solution of carbolic acid with saline.)
3. Into the other 4 tubes pipette 0.5 ml of the carbol saline each.
4. By means of graduated pipette, measure 0.2 ml of the suspected serum into the first tube. Mix well by sucking and blowing out. From this mixture in first tube, transfer 0.5 ml. into the second tube, mix well and from it tansfer 0.5 ml of the mixture into the 3rd tube, mix and then transfer 0.5, ml of this into the 4th and then from this transfer 0.5 ml into the fifth. Mix well and take 0.5 ml from, the fifth tube and discard it. Now the serum is diluted serially starting from the first tube 1 in 5, 1 in 10, 1 in 20, 1 in 40 and 1 in 80.
5. To each of the tubes add 0.5 ml of the antigen, mix well. So now each tube contains 1 ml of the serumantigen mixture and the dilution of the serum is halved i.e. they are now 1 in 10, 1 in 20, 1 in 40, 1 in 80 and 1 in 160.
6. Keep a control tube cantaining 0.5 ml. of carbol-saline and 0.5 ml of the antigen.
7. Keep the rack containing the tubes in a bacteriological incubator (37°C) for 24 hours. After this period, remove the rack and keep at room temperature for 1 hour and then read the result.
8. First examine the control tube and see if it is turbid or has undergone autoagglutination. If auto agglutination has occured, repeat the test. If not then look for agglutination in the five test tubes containing the suspected serum. In a positive case, there should be agglutination in the first 3 tubes at least (i.e. a litre of 1 in 40 at least). In these positive tubes, there will be agglutination of the culture and the supernatant fluid will be transparent

**Note:** The antigen should be stored at 2°C to 5°C temperature. It keeps for 3 months at this temperature.

## DIRECT IMMUNO PEROXIDASE TEST FOR THE DIAGNOSIS OF VIRUS DISEASES

### I. Ammonium Sulphate Precipitation

(a) To 20 ml rabbit anti serum against the specific virus saturated ammonium sulphate (SAS) is added to a final concentration of 40% and the precipitate is washed twice in 40% SAS and finally dissolved in 2 ml of normal saline and dialysed against 0.1M Phosphate Buffer saline (PBS) (pH 6.8) to remove the ammonium sulphate. After estimation of protein, it is directly used for conjugation with the enzyme.

(b) The above immunoglobulin is further purified by gel filtration in Sephadex 0.200 column and the break through peaks are pooled, concentrated by vacuum dialysis and used for conjugation after estimation of protein.

## II. Conjugation of Immunoglobulin with the Enzyme

1. Protein is estimated and to 5 mg of Ig in PBS (pH 6.8) 12 mg of crytalline horse radish peroxidase enzyme is added and 0.05 ml of 1 % aqueous glutaraldehyde is added in drops
2. Conjugation is allowed to take place at room temperature for 2 hours or at 4°C overnight by constant mixing.
3. Excess of glutaraidehyde is removed by dialysis against threee changes of PBS (pH 6.8).
4. The conjugate is centrifuged at 5000 rpm for 10 minutes and diluted 1:10 and stored at 20°C in small aliquotes.

## III. Detection of Rinderpest Viral Antigens in Infected Tissues

1. Buffy coat smears, mouth impression smears and lymph nodes impression smears collected from naturally infected Rinderpest (RP) cases can be used to detect RP viral antigens (Smears from buffy coat of MD cases, brain impression smears from Rabies etc. can be used also for diagnosis of those diseases).
2. The smears and cover slips are fixed in methanol or cold acetone and pre-treated with 10% $H_2O_2$ to remove the endogenous peroxidase activity.
3. The antibody conjugate is added top the smears and incubated at 37°C for 1 1/2 to 3 hours in a humidified chamber.
4. Wash thoroughly in 0.1 M PBS (pH 6.8) to remove excess of conjugate.
5. Add the substrate reactants: freshly mixed (a) 9 ml of benzidine (50 mg/200 ml) + 1 ml of 40% ammonium chloride + 1 ml. of 5% EDIA (pH 6.0) + 2-3 drops of 3% $H_2O_2$.

OR

(b) 0.05% 3'-3' diaminobenzidine tetra chloride in Tris-HCI buffer, pH 7-6 with 0.1% $H_2O_2$

OR

(c) Freshly prepared mixture of 15 mg of Alpha naphthol, 22 mg of P-Phenylne-diamine in 20 ml of 0.07 M PBS, pH 7.2 containing 0.17 M NaCI.

6. Incubate for 5-10 minutes at room temperature and wash in PBS thoroughly.
7. Dry mount and examine under low and high dry lens of a light microscope.
8. Duplicate smears, blocked with untagged antiserum, are used as controls in addition to the normal smears of the corresponding cells and tissues.
9. The cells/tissues containing the viral antigens appear blue or blue black in color.

## TEST FOR BRUCELLOSIS
## MODIFIED ZIEHL — NEELSEN'S STAIN TECHNIQUE
## FOR BRUCELLA

This can be used for demonstration of the Brucella organisms in infected tissues.

**Procedure**

1. Prepare smears from fetal stomach content or placenta and fix by heating.
2. Dilute the corbol—fuchsin used of Acid-fast staining; 1 in 10 and cover the fixed smear with this diluted stain and allow it to act for 15 minutes without heating.
3. The side is decolorised with 0.5% acetic acid solution for 15 seconds and washed thoroughly.
4. Counter stain with Loeffler, methylene blue for 1 minute.
5. Wash, dry and examine under oil immersion.

Brucella stains red color. No other gram negative organisms take this stain.

This stain may be used for demostrating Chlamydia in tissue section.

## THE RAPID TEST FOR BRUCESSOISS WITH COLORED ANTIGEN

The colored Brucella antigen supplied by the IVRI is a suspension of pure smoth cultures of *Brucella abortus* starin 99 in phenolised glycerine saline. The bacteria are colored by crystal violet and brilliant green.

The rapid plate test is useful as a rapid screening procedure and any suspicious case can be confirmed by the tube test.

**The following materials are required for the test**

1. A glass plate, 10" x 8". Ruled on one side, forming one square inch squares. Each square may be separated from its neighbour by double rules.
2. A botle of saline.
3. Small broom sticks
4. A drop bottle containing the colored antigen.

**Procedure**

1. With the pipette drop 0.08 ml of serum in one square and 0.04 in another.
2. If more than one serum is handled, the pipette is rinsed with the saline thrice before the next sample is taken.
3. Add a drop of the antigen from the drop bottle. With the broom stick mix the serum and the antigen quickly. Use separate broom stick for each square, discarding them after mixing is over.
4. Rock the mixture for 4 to 5 minutes with a rotatory movement to ensure thorough mixing. Positive test consists in the aggregation of the antigen into clumps while a

negative test will have a uniform tubidity. Suspicious cases may be confirmed by the tube test. (A sample is suspicious if there is clumping in the 0.08 square and no clumping in the 0.04 square.

**Note:** The antigen should be stored at a temperature of 2°C to 5°C in a refrigerator. It keeps for 9 months.

## ROSE BENGAL PLATE AGGLUTINATION TEST FOR BRUCELLOSIS

This test is similar to the rapid test with the colored antigen described above.

One drop of (0.03 ml) serum is placed on a white enamel tile and one drop of (0.03 ml) of Rose Bengal stained antigen (now available at the Veterinary and Biological Research Institute, Ranipet, Tamil Nadu) is placed next to the serum. The serum antigen mixture is well mixed with a broom stick and the tile rotated for 4 minutes, when the final readings are made and recorded as positive or negative. It is important that the final reading be made at 4 minutes. In a positive test, test, the stained antigen is clumped while the negative test has a uniform turbidity.

In infected herds, this test identifies infected animals at an earlier stage than the serum agglutination test (tube test), and is often positive when the tube test is negative or inconclusive but the complement fixation test is positive.

The antigen that is used for the Card test, that is claimed to be superior to the tube test, is stained with Rose Bengal dye and buffered at pH 3.6.

## TUBE TEST FOR BRUCELLOSIS WITH PLAIN ANTIGEN

The brucella abortus plain antigen supplied by IVRI is a suspension of a pure smooth culture of Brucella abortus strain 99 in phenol saline.

### Procedure

1. Place 5 agglutination tubes in a rack.
2. Pipette 0.8 ml of carbol saline in the first tube and 0.5 ml in the rest of the 4 tubes.
3. Add 0.2 ml of the serum into the first tube, mix well and transfer 0.5 ml of this mixture to the second tube, mix well and then transfer 0.5 ml of the 3rd and so on till the last tube. Discard 0.5 ml from the last tube.
4. Add 0.5 ml of the antigen to each tube and mix well. For mixing a pipette can be used and if only one pipette has to be used for all the 5 tubes, then start from the last tube and then go to the next tube etc. till the first tube is mixed last. The dilutions are 1 in 10 : 1 in 20 : 1 in 40 : 1 in 80 and 1 in 160.
5. A control is made up of 0.5 ml of carbol saline and 0.5ml of the antigen.
6 Incubate the rack in a bacteriological incubator at 37°C for 24 hours. At the end of this period keep the racks at atmospheric temperature for one hour. Then read the results. A positive test consists of agglutination in the first 3 tubes at least. The titre would then be 1 in 40.

**Note:** Store the antigen at 2°C to 5°C temperature. It keeps well for 9 months.

## THE MERCAPTOETHANOL TEST

This test is an agglutination test carried out in the presence of 2-Mercapto ethanol which inactivates IgM molecules present in the serum that is tested. Thus the test may be looked upon as an indicator of the amount, if any, of anti Brucella IgM agglutinin present in the serum. The test is done similar to the tube agglutination test except that the diluent is 0.1 mol/litre 2- mercaptocthanol solution, made as follows:

| | | | |
|---|---|---|---|
| Sodium chloride | — | 8.5 | gms |
| 2-Mercaptoethanol | — | 7.14 | ml |
| Distilled water | — | to 1 | litre |

The folution should be stored at 4°C and prepared fresh every 2-3 weeks. Phenol should not be used at any stage in the mercaptoethanol test.

## BRUCELLA MILK RING TEST

The antigen is known as ABR Antigen and is supplied by IVRI. It consists of a colored suspension of Brucella abortus standardised according to FAO/WHO specifications.

**Procedure**

1. Pool milk from 2 to 5 cows.
2. Measure 2 ml. of this pooled milk into a test tube, 4" x 1/2"
3. Add 2 drops of antigen with a 1 ml paipette that will deliver 16 drops to a ml.
4. Shake the mixture by rotating between the palms. Do not shake vigourously as froth may be produced Mixing should be completed withiun 1 minutes after adding the antigen.
5. Incubate at 37°C for 1 hour.
6. At the end of this period, allow the tubes to stand at room temperature for 90 minutes.

A positive test consists in a cherry colored ring at the cream layer. The antigen and brucella antibodies combine and adhere to the fat and so the ring occurs. But in negative test, no antibodies are formed and so the antigen does not combine with the antibodies and so the antigen is in the skimmed milk portion.

**Note:** If the milk were to be transported to a distance, it can be preserved by adding 2 drops of formalin diluted 1:2.

## DIAGNOSIS OF RABIES

Diagnosis of rabies is confirmed by demonstrating the Negri bodies in the brain.

Most suitable part of the brain is the Ammon's horn or Hippocampus major, which can easily be removed. From this 2 preparations can be made:

(a) An impression smear: This is made by holding a cut end of the Ammon's horn with a forceps and touching this end with a clean slide so that an impression smear is obtained on the slide. A series of 4 or 5 impressioin smears can be made on one slide.

(b) A piece of the brain of Ammon's horn may be placed on a clean slide. With another slide this piece is crushed and a smooth smear is made by drawing the slide with the crushed tissue underneath.

## Staining: Seller's Method

For this method, the smear should be wet.

### Stains

**Stock Solutions:**

| | |
|---|---|
| 1. Methylene blue (color index 922) | 10 gm |
| Acetone free methyl alcohol to make | 1000 ml |
| 2. Basic fuchsin (color index 677) | 5 gm |
| Acetone free methyl alcohol to make | 500 ml |

Keep the stock solutions in screw capped or ground glass stoppered bottles.

**Staining Solution:**

| | |
|---|---|
| Methylene blue solution | 2 Parts |
| Basic fuchsin solution | 1 Part |

Mix well, Keep in a screw cap bottle. prevent from evaporation. Stain keeps indefinitely.

### Procedure

1. Immerse the smears of the brain, *which are still moist* into the staining solution for 1 to 5 seconds.
2. Rinse quickly in running water. Dry in air. Do not use a blotting paper.

Examine under an oil immersion lens. The Negri bodies are bright red and have dark blue to black basophilic inner bodies. All parts of the nerve cells stain blue. The interstitial tissue is pink, Erythrocytes stain a copper hue (orange tinted red).

One can distinguish the Negri bodies from erythrocytes in that the latter are smaller, are copper colored and do not have the dark blue inner bodies.

### Precautions

1. The stains must be kept in ai-tight bottles to prevent evaporation.
2. The smears should be moist

The sucess of straining by Seller's method depends upon the above two factors. So it is not useful for staining smears one may receive by mail. Besides it may not be possible to keep moist smears in hot climate in countries like India. In such places, the following method has been found to be efficacious. (Modified mann's method).

### Stains Required

Make the following stock solutions:

1. Saturated alcoholic solution of Basic fuchsin.

2. Saturated alcoholic solution of Methylene blue.
Keep in stoppered bottles in refrigerator.
From these, make the following solution, just before staining:
Sat. alcoholic solution of Basic fuchsin—2 drops.
Sat. alcoholic solution of Methylene blue—7 drops.
Distilled water. 10ml Mix well.

**Procedure**

1. Make the smear.
2. Fix the smear by keeping the slide in a coplin jar containing methyl alcohol for 15 minutes.
3. Keeping the slide on a staining rack, cover the smear with the freshly made solution.
4. Heat from underneath, till steam rises.
5. Allow the stain for 15 minutes.
6 Wash well, dry in air and examine under oil.

The negri bodies stain bright red while the nerve cells stain blue.

## MUCUS AGGLUTINATION TEST FOR BOVINE VIBRIOSIS

1. Vaginal mucus of the suspected cows are collected by special tampons and quantity of the mucus collected is calculated by the difference in the weight of the tampon befoe and after collection. At least 0.2 gm mucus is collected.
2. The vaginal mucus is extracted from the swab by mixing 5 ml of sterile saline and agitating with glass beads. It is kept overnight at 4°C and centrifuged at high speed for 30 minutes.
3. The supernatant is diluted in doubling series 1:10, 1:20; etc. with 0.3% formal saline.
4. Equal volume of heat killed campylobactor fetus venerealis culture adjusted to an optical density of 0.6 at 525 mu is added to the mucus dilution.
5. Incubate at 50°C overnight and observe for clearnace of the supernatant.
6. Highest dilution of the vaginal mucus giving 50% clearance is considered as the titre.

**Note:** The reaction is mainly due to IgA in covalescent animals.

## AGGLUTINATION TEST FOR AVIAN MYCOPLASMOSIS

### A. RAPID PLATE TEST

Mycoplasm gallisepticum colored antigen stained with Rose Bengal dye is available from the Institute of veterinary preventive Medicine, Ranipet.

**Procedure**

A drop (0.03 ml) of suspected serum and an equal quantity of the colored antigen are placed on a glass plate and mixed with a broom sick. Carefully obserrve for agglutination, if any. For controls known positive and negative sera should be used.

In a positive case agglutination of the stained particles into clumps occurs within 1 minute.

In a negative reaction, no agglutination occurs.

In non-specific reactions fine agglunination with no clear interspace appears after 2 minutes.

**Note:** Serum to be tested should be examined within 3 days of collections. It should be non-hemolysed and should not be frozen or contaminated.

## B. TUBE AGGLUTINATION TEST

1. The sera sample are serialiy diluted (double) 1 : 12.5; 1 : 25; 1 : 50; 1 : 100; and 1 : 200, in phosphate buffer saline (pH 7.2).
2. An equal volume of the colored *M. gallisepticum* antigen that is used for plate test is added to the serum dilutions.
3. Incubate at 37° C for 24 to 48 hrs.
4. Clumping of the antigen is considered as the end point.
5. Highest dilutions of the serum agglutinating the antigen is the titre.
6. Any titre 1: 50 and above is considered positive.
7. Controls with known positive and negative sera should be used.

## FLUORESCENT ANTIBODY TECHNIQUE

Purified antibody tagged with a fluorochrome when exposed to U.V. rays gives rise to specific fluorescence. The test is based on the antigen-antibody reaction.

**Materials Required:** Hyper-immune serum, fluorescent-iso-thiocyanate, ammonium sulphate, sephadex-G-25, Dialysis membrane, PBS.

(a) High titre serum is precipitated with Saturated Ammonium sulphate solution to purify the antibody. Protein is estimated. It should contain atleast 25 mg/ml of protein.

(b) Conjugation with FITC: (i.) The fluorochrome is dissolved in 0.15 M $Na_2HPO_4, 2H_2O$ (pH9.0) at the rate of 1 mg/ml. (ii) 1 ml of dye solution is added for every 100 mg of protein. Stirr at room temperature. (iii) pH is adjusted to 9.5 by the addition of 0.1 M $Na_3 PO_4 . 10 H_2O$ while stirring. The reaction is allowed to continue for 60 min. the pH being maintained at 9.5.

(c) The ree dye is removed by passing through a Sephadex-G-25.

(d) **Staining Technique:** (i) The suspected smear is made on a clean slide, fixed in acetone, and washed in PBS (0.1 M, ph 7.2) (ii) Add a drop of FITC-ab conjugate on the smear, incubate at 37°C for 1-3 hr in a humidified chamber, (iii) Wash thoroughly in PBS.

(e) **Examination of the smears:** (i) Add a drop of 90% glycerin in PBS to the condenser of the fluorescent Microscope. (ii) Place slide on the stage such that the drop of glycerine touches the slide. (iii) Examine under a fluorescent microscope using HBO 200 bulb BG. 12 exciting filter and yellow interference filter. (iv) Positive smears will give apple green fluorescence.

## SURFACE FIXATION TEST FOR LISTERIOSIS

Since Gram positive bacteria tend to autoagglutinate the agglutination test is not useful. Surface fixation test was first introduced by Casteneda for the detection of Brucella antibodies.

**Procedure**

(1) A coloured antigen suspension is prepared using killed Listeria monocytogenes bacteria using crystal violet.

(2) Filter paper strips of 1 cm x 5 cm are in Whatmann paper No. 42.

(3) A small circle is marked at one end of the filter paper and a drop of suspected antiserum is place within the circle and allowed to dry.

(4) Similar strip is prepared using a normal serum of the species which is being tested as the control.

(5) A drop of coloured antigen is placed over the dried serum-spots on the filter paper strips and allowed to dry.

(6) Now dip the filter paper strips into normal saline contained in a beaker such that only the tip ( nearer the circle) just touches the surface of the saline.

(7) Immediately, the saline starts moving upwards due to capillary action. If the serum sample is positive for antibodies the coloured antigen-antibody spot remians within the circle. In the case of negative sera and in the control the coloured antigen is not fixed and hence moves along the saline forming a trail.

APPENDIX 1

# A SUMMARY OF CAUSES OF INFERTILITY IN COWS

(Contributed by Dr P Narasima Rao)

## A. INFECTIOUS AGENTS

I. Venereal infections :

1. Trichomoniasis
2. Vibriuosis
3. Brucellosis
4. Granular vaginitis
5. Vesicular vaginitis
6. Epivag (Viral, in south Africa)

II. Miscellaneous general infections:

1. Tuberculosis
2. Actinomycosis
3. Pleuropneumonia-like organisms
4. Stereptococcal infections
5. Corynebacterium pyogenes
6. Infections Bovine Rhino-tracheitis

## B. PHYSIOLOGICAL CAUSES

Hormonal Disturbaces:

1. *Cystic ovaries*
2. *Failure of heat.*

**(a) Anestrum associated with corpus luteum**

(i) Anestrum due to pregnancy.

(ii) Silent heat due to low levels of estrogen or higher estrogen threshold for some animals.

(iii) Due to unobserved heat. (Owner's ignorance)
(iv) Associated with uterine pathology: pyometra fetal maceration, mummified fetus and endometritis associated with failure of regression of corpus luteum.
(v) Anestrus without uterine disease: corpus luteum of previous heat or of previous parturition. persists.
(vi) Retained corpus luteum associated with embryonic death.
(vii) Cystic corpus luteum.

**(b) Anestrum without a corpus luterm**

(i) Small inactive ovaries—due to lack of FSH
(ii) Cystic ovaries—failure of ovulation due to lack or insuffcient LH release.
(iii) Inanition—old age, loss of teeth, severe mange.
(iv) Chronic disease—causing infertility — Johne's Disease, Actiuonmycosis, Tuberculosis, after effects of Foot and Mouth Disease, heavy parasitic infections, chronic acetonemia—these cause debility and so anestrus.
(v) Seasonal variation—decrease in length of day time. too high or too low temperature causes anestrus. Especially in cross-bred animals and buffaloes.
(vi) Miscellaneous conditions—stress of lactation, freemartinism, congenital bilateral hypoplasia of ovaries tumors of ovaries.

**II Other Causes of Physiological Infertility Failure of Fertillsation**

(i) Obstruction of oviduct—adhesion of ovarian bursa, chronic salpingitis, pyo and hydro-salpinx developmental anomalies of oviduct.
(ii) Abnormalities in ovulation—failure of ovulation and delayed ovula-tion
(iii) Multiple ovulation.
(iv) Inability of ova to be fertilised—degeneration of ovum and ageing of ovum.
(v) Inability of the sperm to fertilise—lack of capacitation, ageing of spermatozoa, morphological defects of spermatozoa.
(vi) Early embryonic death—due to sex steroid Imbalance endometritis, genetic.

## C. HERDITARY FORMS OF INFERTILITY

### I. Anatomical Defects

(i) Hypoplasia of ovaries—due to failure of migration of Primordial germ cells.
(ii) White Heifer's Disease—imperforate hymn—due to a single recessive sex linked gene, linked with gens for white color. Vagina may be completely absent.
(iii) Abnormal wolffian or garnter's ducts—A cul de sac is formed on the ventral wall of the vagina.
(iv) Double external os of cervix—hertiable.
(v) Freemartinism.
(vi) Absence of gonads—dominant autosomal factor.
(vii) Cystic ovaries or nymphomania.

(viii) Infantile vulva.

## II. Miscellaneous Hereditary Forms of Infertillity

(i) Abnormal sex behaviour—variation in length of estrus cycle, time of occurrence of postpartuient heat, time of ovulation, intensity of signs and heat appear to be inherited
(ii) Inbreeding reduces fertility.
(iii) Embryonic mortality—may be inherited: may be due to adverse maternal uterine environment.
(iv) Inheritance of infertility.

## D. PATHOLOGICAL CAUSES

**I. Ovary**
(a) Oophoritis. (b) Parovarian cysts. (c) Neoplasms.

**II. Oviduct**
(a) Ovarian adhesions (b) Hydrosalpinx (c) Pyosalpinx.

**III. Uterus**
(a) endometritis (b) Pyometra (c) mucometra
(d) Septic metritis (e) Sclerotic metritis (f) Delayed involution
(g) Perimetritis (h) abscess of the uterine wall
(i) Cystic degeneration of the uterine wall (j) Uterine abscess
(k) Tumors

**IV. Cervix**
(a) Cervicitis (b) Stenosis of cervix (c) Cysts of cervix
(d) Mucocervix (e) Tumors

**V. Vagina**
(a) Vaginits (b) Cysts of Vagina (c) Tumors

**VI. Vulva**
(a) Vulvitis (b) Tumors

## E. MANAGEMENTAL AND NUTRITIONAL CAUSES OF STERILITY

I. Faliure of observation of heat — Mostly in buffaloes, in which it is silent.
II. Insemination during early or late heat.
III. Improper A.I. Technique.
IV. Lack of energy and mineral requirements
High energy ratio is required for regular breeding. Proper balance in Ca : P is essential for quick involution of the uterus and for the animal to come into

APPENDIX II

# MATERIALS TO BE SENT FOR LABORATORY ANALYSIS

(in cases of Nutritioanl Disorders)

Diseases of nutritional origin in animals may be due to deficiency of a particular nutrient or nutrients in the rations or may be due to presence of interfering factors or toxic principles. For investigation and diagnosis of such conditons analysis of feeds, fodders and biological materials has to be carried out. Materials to be sent and their method of despatch are given below:

| | Condition | Materials to be Sent | Quantity |
|---|---|---|---|
| 1. | Protein deficiency | (a) Serum<br>(b) Oxalated blood | 5 ml with a drop of toluene.<br>10 ml. |
| 2. | Serum Calcium, Phosphorus and Magnesium deficiency | Serum | 5 ml with a drop of toluene (Do not use Ammonium Oxalate) |
| 3. | Copper deficiency | Oxalated blood | 10 ml |
| 4. | Iodine deficiency | Serum | 20 ml |
| 5. | Fluorosis | (1) Water<br>(2) Urine | 500 ml<br>100 ml with a drop of toulene |
| 6. | Vitamin A estimation | Serum | 15 ml |
| 7. | Estimation of serum alkaline phosphatase | Serum | 2 ml |
| 8. | For chemical analysis of rations | (a) Concentrates<br>(b) Concentrate mixture<br>(c) Feeds and fodders | 500 gms each packed in paper |

APPENDIX III

# COLLECTION OF SPECIMENS FOR LABORATORY EXAMINATION

**For Bacteriological Examination**

Intestines: Ligate about 3 inches of the bowel at either ends and send the loop unopened for bacteriological examination of the intestinal flora, at once (in ice if it is to be transported for some distance).

**Heart blood swabs:** For this you require sterile swabs, sterile scalpel, sterile forceps and a spatula. Before opening the chambers, singe the epicardium over the right ventricle with the hot spatula (this is heated on a spirit lamp). At the singed place, make an incision with the sterile sealpel, dilate the opening with the sterile forceps and then insert the sterile swab into the incision and collect the blood in the swab into the tube containing the charcoal transport medium, carefully seal it amd forward to the laboratory.

**Charcoal Transport Medium Composition:**

| | |
|---|---|
| Sodium chloride | 4.00 gms |
| Potassium chloride | 0.20 gm |
| Dipotassium phosphate | 1.70 gms |
| Charcoal (bacteriological) | 10.00 gms |
| Agar | 4.00 gms |

Dissolve the salts in 500 ml of distilled water and then add charcool. Add the agar separately in 500 ml distilled water and boil. After it is dissolved, mix the two solutions, autoclave at 121°C at 15 lbs for 30 minutes. Dispense 8 ml of the semi-solid medium in screwcapped tubes and store at 4°C. This medium is useful for sending materials for isolation of viruses also.

**Utertus:** If there are abnormal contents in the uterus, a small segment is isolated between ligatures and retained unopened for bacteroiological examination.

**For histopathological examination:** For histopathological examination to be of any use, the tissue should be fresh without purefaction or postmortem autolysis. So the carcase should be opened immediately after death and the tissue fixed immediately after removing from the carcase.

Keep always wide mouthed, stoppered bottles of 8 ounce capacity, three quarters filled with 10% formol-salime. (This is prepared by adding 10 ml of formalin supplied by the Medical Stores Depot to 90 ml of normal saline). Into these bottles drop the tissue immediately, when it is being examined. The tissue need not be larger than 1 cm x 1 cm x 1 cm. If whole organs have to be preserved, large containers must be used.

**For examination of parasites:** Do not handle the parasites, especially the ascarides, with bare hands. As soon as the parasite is collected from the carcase put it in hot 5% formalin for 2 minutes. Then transfer them into cold 10% tormalin opr 70% alcohol to which a little glycerine is added and then forward for the laboratory.

## Procedure to be Observed in Collecting Materials from Suspected Cases of Poisoning for Chemical Analysis

The entire alimentary canal should be opened and its contents inspected for suspicious looking substances. If any suspicious looking substances are detected in the alimentary canal,they should be packed in a separate vessel, and spirit should not be added unless necessary for its preservation.

The following articles should be sent from horses and cattle for examination:

(a) About two pounds of the contents of the stomach, with about a pound of contents of the intestines in a clean glass or well glazed earthen vessel or vessels with the addition of sufficient quantity of strong unmethylated spirit for their preservation.

(b) (i) About a pound of the liver and a similar weight of the stomach in a separate clean glass or a well-glazed earthen vessel with the addition of sufficient quantity of strong uemethylated spirit (95% alcohol or rectified spirit).

(ii) The amount of spirit added to each vessel should be measured and the amount so added written on the lable affixed to the vessel.

(iii) Methylated spirit supplied by the Medical Stores Depot, may however, be used for the purpose if unmethylated spirit is not available.

(c) Dry dung without the addition of spirit.

(d) Any suspected poisons and

(e) At least 4 oz, of the sample of the spirit used in preserving the articles.

All bottles and packets should be carefully sealed by the officer making the examination, and closed in such a manner that they cannot be opened without destroying the seal.

When an officer forwards articales to the Chemical Examiner for examination, he should at the same time address and foward separately a letter to the Chemical examiner advising their despatch. This letter should contain:

(a) impression of the seal used in closing the vessel and a description there of;

(b) a list of articles forwarded and information as to how the articles have been forwarded;

(c) the name of the officer from whom the order has been received to forward the articles, and the number and dete of such order;

(d) information as to the number and kind of animals affected and number of deaths;

(e) any information obtainable as to postmortem appearance, nature ane duration of

APPENDIX IV

# MATERIALS TO BE SUBMITTED TO THE LABORATORY FOR DIAGNOSIS OF A BACTERIAL DISEASES

| Disease | Materials to be Submitted |
|---|---|
| Actinobacilliosis | 1) Smears of pus from lesions.<br>2) Formalin preserved material from lestions (affected muscle), |
| Actinomycosis | — do — |
| Anthax | 1) Blood smear from the tip of ear or tail.<br>2) Smear from swelling.<br>3) Blood swab for cultural exam.<br>4) Muzzle piece for biological test, in boric acid. |
| Bacillary White Diarrhoea | 1) Dead chick.<br>2) Serum sample from adult bird. |
| Black disease | Send piece of liver in sterile container. |
| Black quarter | 1) Smears from the swelling.<br>2) Dry a piece of affected muscle and send in sterile container packed in salt. |

| Disease | Materials to be Submitted |
|---|---|
| Bovine lymphangitis | 1) Smears from gland.<br>2) Pus from lesions in charcoal medium. |
| Botryomycosis | 1) Pus smears.<br>2) Piece of tissue in 10% formol-saline. |
| Brucellosis | 1) Collect serum 3 weeks after abortion and send.<br>2) Fetal stomach content in charcoal medium.<br>3) Swab from uterine discharge in charcoal medium.<br>4) Milk in sterile container. |
| Chronic Respiratory Disease | 1) Trachea on ice.<br>2) Serum. |
| Contagious bovine pleuropneumonia | 1) Formalin preserved pulmonary lesion.<br>2) Serum 3) Piece of lung in ice. |
| Contagious Caprine pleuropneumonia | 1) Fornalin preserved pulmonary lesion.<br>2) Piece of lung in ice. |
| Enterotoxemea | Send a loop of 12" bowel preserved in 0.5% chloroform or in ice. |
| Epizootic lyphangitis | 1) Smears from the pus.<br>2) Affected lymph nodes in ice. |
| Fowl cholera | 1) Blood smears.<br>2) Blood swabs in charcoal medium.<br>3) Long bone packed in charcoal. |
| Fowl spirochaetosis | 1) Blood smears.<br>2) Pieces of liver and spleen preserved in formalin. |
| Fowl typhoid | Liver in formalin. |
| Glanders | 1) Smears from discharge.<br>2) Lesions preserved in 10% formalin-saline. Note I<br>**This disease comes under Glanders and Farcy act** |
| Hemorrhagic septicema | 1) Blood smear<br>2) Blood swabsin charcoal medium.<br>3) Smear from edematous fluid. |
| Johne'disease | 1) Backrake and send bowel washing in a sterile bottle.<br>2) Make a smear from the mucous pinched from the rectum, fix heat and send.<br>3) Send affected piece of intestine and mesenteric lymphatic gland in 10% formalin-salin. |
| Leptospirosis | 1) Blood serum taken 3 weeks following abortion.<br>2) Pieces of liver and kidney preserved in formalin.<br>3) Milk or urine in vials. Add 1 drop of formalin to 20ml<br>4) Blood in vial. |
| Listerosis | 1) Send brain preserved in ice or charcoal medium.<br>2) Send formalin preserved brain. |

| Disease | Materials to be Submitted |
|---|---|
| Strangles | 1) Pus smear and<br>2) Swab in charcoal medium. |
| Swine erysipelas | Send blood for isolation of organic in acute stage I or sections of spleen, kidney, liver, packed in ice or charcoal medium. |
| Tuberculosis | 1) Suspected lesionsin 10% formol-saline.<br>2) Milk in sterile containers,<br>3) Pus from lesions. |
| Ulcerative lymphangtis | 1) Smears and 2) swabs from pus in charcoal medium |
| | **B. VIRAL DISEASES** |
| African Horse sickness | 1) Serum from recovered cases, which are rare.<br>2) Brain, spleen on ice or in charcoal medium for isolation of the virus. |
| Avian encephalo-myelitis | Brain and spinal cord in 50% glycerine. |
| Canine distemper | Pieces of lungs, bladder, liver, trachea, stomach wall and brain in 10% formol-saline. |
| Equine encephalo-myelitis | 1) Brain in 10% formol-saline and in ice or charcoal medium.<br>2) Serum. |
| Equine infectious anemia | 1) Serum for albumin; globulin ratio.<br>2) Whole blood for animal inoculation. |
| Foot and Mouth disease | Vesicular fluid and curretted epithelium in 50% glycerine phosphate buffer. |
| Hog cholera | 1) Serum from affected or ailing animals.<br>2) Pieces of spleen, liver and kidney in charcoal medium. |
| Infectious bovine rhinotracheitis | 1) Affected trachea and turbinate bones in 10% formol-saline.<br>2) Serum samples for in serum-neutralisation test. |
| Infectious bronchitis | 1) Trachea and bronchi in charcoal medium.<br>2) Serum from recovered chicks. |
| Infectious bursal disease(Gumboro-disease) | 1) Serum<br>2) Bursa fabricius and kidney in charcoal medium.<br>3) Bursa Fabricius and kidney in 10% formal-saline. |
| Infectious canine hepatitis Liver. | Gall bladder and kidney in 10% formol-saline. |
| Infectious laryngo-tracheitis | 1) Trachea in charcoal medium.<br>2) Smears from tracheal mucosa.<br>3) Serum samples from recovered birds. |

| Disease | Materials to be Submitted |
|---|---|
| Marek's Disease | 1) Live affected birds for collection of materials for isolation of virus and histopathology.<br>2) Serum,<br>3) Suspected materials in 10% formol-saline and charcoal medium. |
| Ornithosis and Psittacosis | 1) Spleen and kidney in charcoal medium.<br>2) Impression smears from organs and smears of exudates. |
| Pox | Scabs and vesicular fluid in charcoal medium. |
| Rabies | 1) Send half of the brain in 50% glycerine.<br>2) The other half in dichromate solution made as follows; Potassium dichromete 3 G. glacial acetic acid 5 ml. distilled water 95 ml. |
| Ranikhet disease | 1) Send liver and spleen in 50% glycerine.<br>2) Proventriculus in 10% formol-saline. |
| Rinderpest | 1) Intestinal lesions in 10% formol saline.<br>2) Pieces of spleen and lymphnode in ice and in charcoal medium<br>3) Impression smears from buccal mucosa.<br>4) Defibrinated or citrated blood. |
| Scrapie | Brain in 10% formol-saline. |

### C. FUNGAL DISEASES

| Disease | Materials to be Submitted |
|---|---|
| Ringworm | Skin scrapings in an envelope. |
| Aspergillosis | Pieces of affected lung in 10% formol-saline. |
| Coccidioidomycosis | Pleces of lung and lymph gland in 10% formol-saline. |
| Rhinosporidiosis | 1) Nasal discharge in a bottle preserved with a drop of toluene.<br>2) Piece of the growth in 10% formol-saline. |
| Mucormycosis | Pieces of affected tissues in 10% formol-saline. |
| Histoplasmosis | 1) -ditto- |
| Aflatoxicosis | 1) Suspected feed (groundnut cake, especially), 500 gms. well packed.<br>2) Liver piece in 10% formol-saline. |

### D. PROTOZOAL DISEASES

| Disease | Materials to be Submitted |
|---|---|
| Coccidiosis | 1) Feces and scrapings of the intestines preserved in 2% potassium dichromate solution or in 1% formalin. |
| Toxoplasmosis | 1) Exudate in 1 in 10000 merthiolate.<br>2) Lesions in 10% formol-saline. |
| Babeslosis and Surra | Blood smears |
| Theileriasis | 1) Blood smears<br>2) Perform a biopsy on prescapular lymph gland and make a |

| Disease | Materials to be Submitted |
|---|---|
| | smear of the pulp obtained. Fix the smear in methyl alcohol for 15 minutes and then send the smears. |
| | 3) From a dead animal send lymph gland and spleen in 10% formol-saline. |
| Anaplasmosis | Blood smears. |